MW00982303

Z.-Q. Liu, J. Cai, R. Buse

Handwriting Recognition

Springer

Berlin
Heidelberg
New York
Hong Kong
London
Milano
Paris
Tokyo

Studies in Fuzziness and Soft Computing, Volume 133
http://www.springer.de/cgi-bin/search_book.pl?series=2941

Further volumes of this series can be found on our homepage

Vol. 114. M. Fitting and E. Orowska (Eds.)
Beyond Two: Theory and Applications of Multiple Valued Logic, 2003
ISBN 3-7908-1541-1

Vol. 115. J.J. Buckley
Fuzzy Probabilities, 2003
ISBN 3-7908-1542-X

Vol. 116. C. Zhou, D. Maravall and D. Ruan (Eds.)
Autonomous Robotic Systems, 2003
ISBN 3-7908-1546-2

Vol 117. O. Castillo, P. Melin
Soft Computing and Fractal Theory for Intelligent Manufacturing, 2003
ISBN 3-7908-1547-0

Vol. 118. M. Wygralak
Cardinalities of Fuzzy Sets, 2003
ISBN 3-540-00337-1

Vol. 119. Karmeshu (Ed.)
Entropy Measures, Maximum Entropy Principle and Emerging Applications, 2003
ISBN 3-540-00242-1

Vol. 120. H.M. Cartwright, L.M. Sztandera (Eds.)
Soft Computing Approaches in Chemistry, 2003
ISBN 3-540-00245-6

Vol. 121. J. Lee (Ed.)
Software Engineering with Computational Intelligence, 2003
ISBN 3-540-00472-6

Vol. 122. M. Nachtegael, D. Van der Weken, D. Van de Ville and E.E. Kerre (Eds.)
Fuzzy Filters for Image Processing, 2003
ISBN 3-540-00465-3

Vol. 123. V. Torra (Ed.)
Information Fusion in Data Mining, 2003
ISBN 3-540-00676-1

Vol. 124. X. Yu, J. Kacprzyk (Eds.)
Applied Decision Support with Soft Computing, 2003
ISBN 3-540-02491-3

Vol. 125. M. Inuiguchi, S. Hirano and S. Tsumoto (Eds.)
Rough Set Theory and Granular Computing, 2003
ISBN 3-540-00574-9

Vol. 126. J.-L. Verdegay (Ed.)
Fuzzy Sets Based Heuristics for Optimization, 2003
ISBN 3-540-00551-X

Vol 127. L. Reznik, V. Kreinovich (Eds.)
Soft Computing in Measurement and Information Acquisition, 2003
ISBN 3-540-00246-4

Vol 128. J. Casillas, O. Cordón, F. Herrera, L. Magdalena (Eds.)
Interpretability Issues in Fuzzy Modeling, 2003
ISBN 3-540-02932-X

Vol 129. J. Casillas, O. Cordón, F. Herrera, L. Magdalena (Eds.)
Accuracy Improvements in Linguistic Fuzzy Modeling, 2003
ISBN 3-540-02933-8

Vol 130. P.S. Nair
Uncertainty in Multi-Source Databases, 2003
ISBN 3-540-03242-8

Vol 131. J.N. Mordeson, D.S. Malik, N. Kuroki
Fuzzy Semigroups, 2003
ISBN 3-540-03243-6

Vol 132. Y. Xu, D. Ruan, K. Qin, J. Liu
Lattice-Valued Logic, 2003
ISBN 3-540-40175-X

Zhi-Qiang Liu
Jinhai Cai
Richard Buse

Handwriting Recognition

Soft Computing and Probabilistic Approaches

Springer

Professor Dr. Zhi-Qiang Liu
City University of Hong Kong
School of Creative Media
Tat Chee Avenue 83
Kowloon, Hong Kong
PR China
E-mail: smzliu@cityu.edu.hk

and

Department of Computer Science
and Software Engineering
The University of Melbourne
Victoria 3010
Australia
E-mail: zliu@csse.melbourne.edu

Dr. Jinhai Cai
School of Software Engineering
& Data Communications
Queensland University of Technology
GPO Box 2434
Brisbane QLD 4001
Australia
E-mail: j.cai@qut.edu.au

Richard Buse
Department of Computer Science
and Software Engineering
The University of Melbourne
Victoria 3010
Australia

ISSN 1434-9922
ISBN 3-540-40177-6 Springer-Verlag Berlin Heidelberg New York

Library of Congress Cataloging-in-Publication-Data applied for

A catalog record for this book is available from the Library of Congress.

Bibliographic information published by Die Deutsche Bibliothek
Die Deutsche Bibliothek lists this publication in the Deutsche Nationalbibliographie; detailed bibliographic data is available in the internet at <http://dnb.ddb.de>.

Springer-Verlag Berlin Heidelberg New York
a member of BertelsmannSpringer Science+Business Media GmbH
http://www.springer.de

Printed in Germany

Typesetting: camera-ready by authors
Cover design: E. Kirchner, Springer-Verlag, Heidelberg
Printed on acid free paper 62/3020/M - 5 4 3 2 1 0

Abstract

Off-line handwriting recognition systems can be broadly divided into three categories: statistical, syntactic, and soft computing-based approaches. Most current systems use traditional statistical approaches that make decisions based on the statistical information present in data. This book takes a fresh look at the problem of unconstrained handwriting recognition and introduces to the reader new techniques using statistical and soft computing approaches. The text discusses in detail the types of uncertainties and variations present in handwriting data. Since handwritings are 2-D data, this book presents several algorithms that use modified hidden Markov models and Markov random field models to model the handwriting data statistically and structurally in a *single* framework. As it is well-known that many uncertainties in handwritings cannot be modeled adequately by traditional statistical methods, the recognition of a word in different styles may be best accomplished by fuzzy logic. The book explores methods that use fuzzy logic and fuzzy sets for handwriting recognition. The effectiveness of these techniques is demonstrated through extensive experimental results on real handwritten characters and words.

Acknowledgments

The authors would like to thank the staff of the Department of Computer Science and Software Engineering, the University of Melbourne for their generous support and members of the Computer Vision and Machine Intelligence Laboratory for many scholarly debates and inspiring discussions. Zhi-Qiang Liu would like to thank the School of Creative Media, City University of Hong Kong for its support during the preparation of the final draft of this book. Jin-Hai Cai is grateful for all the people in the School of Software Engineering and Data Communications, Queensland University of Technology for their help. We would also like to thank Yee-Cheng Ng for her proofreading of Chapter 4 of the book. Finally, we are very much obliged to Helen Chu Lee for her meticulous proofreading of the entire manuscript. It is her tireless effort that has greatly improved the book.

Preface

Over the last few decades, research on handwriting recognition has made impressive progress. The research and development on handwritten word recognition are to a large degree motivated by many application areas, such as automated postal address and code reading, data acquisition in banks, text-voice conversion, security, etc. As the prices of scanners, computers and handwriting-input devices are falling steadily, we have seen an increased demand for handwriting recognition systems and software packages. Some commercial handwriting recognition systems are now available in the market. Current commercial systems have an impressive performance in recognizing machine-printed characters and neatly written texts. For instance, High-Tech Solutions in Israel has developed several products for container ID recognition, car license plate recognition and package label recognition. Xerox in the U.S. has developed TextBridge for converting hardcopy documents into electronic document files.

In spite of the impressive progress, there is still a significant performance gap between the human and the machine in recognizing off-line unconstrained handwritten characters and words. The difficulties encountered in recognizing unconstrained handwritings are mainly caused by huge variations in writing styles and the overlapping and the interconnection of neighboring characters. Furthermore, many applications demand very high recognition accuracy and reliability. For example, in the banking sector, although automated teller machines (ATMs) and networked banking systems are now widely available, many transactions are still carried out in the form of cheques. To integrate conventional and e-business models and save operational costs, it is desirable to have automated systems that are able to recognize, interpret, and verify what has been written on the cheque and carry out the required transactions, e.g., bill payments, deposits, and so on. In such applications, the handwriting recognition system must be able to perform *at least* as well as the bank teller; otherwise it will result in too many returned cheques or transaction errors, which would be disastrous in some cases.

In general, a distinction is made between on-line and off-line script in handwriting recognition. For on-line systems, the dynamic (temporal) information about the writing speed and acceleration, and about the order of line-segments making up a word or character is available to the recognizer. That is, in on-line systems we may use spatial-temporal information. As a result, on-line handwriting recognition systems often have higher recognition rates than that of their off-line counterparts. Several well-established algorithms, such as hidden Markov models and dynamic programming, can be applied readily in the on-line case.

In off-line systems, variations in handwritings may result not only from different writing styles but also from different writing media such as pen and paper, etc. Broadly, an off-line handwriting recognition system includes three parts: image pre-processing, feature extractor, and classifier. Pre-processing is primarily used to reduce variations of handwritten characters. It usually involves noise reduction, slant correction, and size normalization. Feature extractor is essential for efficient data representation and extracting, hopefully, most discriminant features to be used in the recognition process. The classifier makes a final decision according to the features and some additional knowledge. In general, off-line handwriting recognition approaches can be divided into two basic categories:

- Segmentation-based approaches that first segment the word into characters or sub-character parts. The recognition process is then accomplished with the aid of a dictionary. Many segmentation algorithms are available in the literature [26, 48, 43, 166]. However, characters in unconstrained handwritten words often connect or even overlap with neighboring characters, which makes it difficult to tell where one character ends and another begins. As a result, such approaches are susceptible to segmentation errors.

- Word-based (holistic) approaches that use no segmentation. Such systems do not rely on individual characters, instead, they recognize words as entities [82, 192, 228]. Although these systems do not require segmentation, they need at least one model or template for each word.

In terms of recognition techniques, off-line handwriting recognition systems can also be divided into three main categories: statistical, syntactic, and soft computing-based approaches. Statistical approaches make a decision by statistical information derived from the input data. In most systems, the probability density functions are assumed to be Gaussian functions or Gaussian mixtures mainly because they are mathematically tractable.[1] Approaches in this category are relatively insensitive to noise and distortions caused by the feature extraction process; however, it is difficult to use statistical model to describe structural information. On the other hand, syntactic approaches model handwritten images by structures that are composed of primitive patterns according to some structural descriptions or a set of rules. There are two major disadvantages in these approaches: First, it is difficult to include statistical information present in the handwriting in the

[1]Although most traditionalists would insist that the Gaussian model represents the data distribution, the lack of robustness in real applications demonstrates that such an assumption has little relevance to reality.

recognition process. This may result in a recognition system that is sensitive to noise. Second, the syntactic descriptions may not able to model distortions in data, such as broken lines, filled holes, etc. which unfortunately happen rather frequently in handwriting images. Neural network approaches have become popular in handwriting recognition since the 1980s. This is mainly attributed to neural network's powerful learning capability and the flexibility of input features. Recently, genetic algorithms have been adopted in handwriting recognition by many researchers. Genetic algorithms are stochastic global search methods, which mimic the mechanism of natural biological evolution. More recently researchers have developed soft computing techniques for handwriting recognition and received encouraging results.

In this book, we present several new techniques for the recognition of handwritten words and characters using both statistical and soft computing approaches. The book is organized as follows:

Chapter 1 introduces the general area of handwritten character recognition, discusses major problems and techniques available in the literature, and overviews the state-of-the-art research and development. More specifically, we discuss some major techniques for feature extraction, which are divided broadly into two categories: features from binary images and from gray-scale images. We will present some of the most popular and important classification methods for handwriting recognition, and discuss their properties.

In Chapter 2, we present pre-processing techniques for slant and correction, size normalization, and baseline determination. We also present in detail the use of the Gabor filter in feature extraction. The most prominent merits of the Gabor filter-based feature extraction method are that it is robust to noise and is able to extract features from both binary images and gray-scale images. In order to extract useful features regardless of the average line-width of handwritten characters or words, we present an algorithm for adaptive estimation of the parameters of Gabor filters. We also introduce several new techniques for extracting such features as skeletons, outer contours, and oriented line segments used in handwriting recognition.

Chapters 3 to 5 present Markov model-based approaches to recognizing handwritten numerals. Chapter 3 describes a new technique for recognizing totally unconstrained handwritten numerals based on hidden Markov models. This approach integrates the structural and statistical information to make a decision. The statistical information is modeled using hidden Markov models, where state-duration-adapted transition probabilities are used to improved the modeling of state durations in the conventional hidden Markov models. To improve the ability of statistical methods in modeling pattern structures, we present the concept of *macro states* for capturing the structural information. In order to speed up the recognition process, we present a modified Viterbi algorithm to avoid the logarithmic and exponential operations that are computationally intensive. The experimental results show that this approach can achieve high performance in terms of speed and accuracy. In Chapter 4 we present a method using a 2-D hidden Markov process to model spectral features for recognizing handwritten numerals. Because Fourier descriptors can achieve invariance to rotation, reflection and scale, they have been widely used for recognizing 2-D shapes such as tools and handwritten numerals. This chapter demonstrates that the 2-D hidden Markov models with features extracted using Fourier descriptors can obtain good results. Since handwritings are 2-D data, we may use Markov random field (MRF) models to model handwritings in their natural data format. In this way, we are able to model the handwriting data statistically and structurally in a single framework. Chapter 5 discusses the important issues of Markov random field models including the framework of MRFs, the neighborhood systems, the design of clique functions and calculation of the global optimum. We apply MRFs to recognizing handwritten numerals.

Chapters 6 to 8 present new techniques for handwritten word recognition. In Chapter 6, we present a method using Markov random field model for unconstrained handwritten word recognition. Again we use the Gabor filter to extract directional features. We use the MRFs to model the relationships between line-segments that have similar orientations within the neighborhood system. Based on these relationships, we define a similarity measure between the templates and images. We establish fuzzy neighborhood systems and design fuzzy matching measures to handle large variations in the data. We use relaxation labeling to maximize the global compatibilities of the MRF models. This chapter also discusses the influence of neighborhood sizes and iteration number of relaxation labeling on recognition rates. Chapter 7 describes a new off-line word recognition system to recognize unconstrained handwritten words based on structural and relational information in handwritten words. We use Gabor filters to extract features from words, and then use an evidence-based approach for word classification. We also present a new method for estimating the Gabor filter parameters, enabling the Gabor filter to be automatically tuned to the word image properties. We introduce two new methods for correcting the slant and tilt present in handwritten words. Throughout the chapters, we present some experimental results.

When writing a given word, say, "*greetings*" there may be large variations between the written words by different people. However, after preprocessing, these different versions of "*greetings*" share some *core* common features that provide the essential clues for recognition. In human perception, such core features play a critical role in identifying the word written in different styles. We argue that the majority of writing variations cannot be modeled adequately by statistical methods and that the recognition of the word in different styles is a process of gathering the *degree* (fuzzy) of evidence from the core features to make a decision. In Chapter 8, we present an off-line word recognition system using fuzzy logic. This system uses structural information in the unconstrained written word. Oriented features in the word are extracted with the Gabor filter. Based on the oriented features, we associate each word with a set of fuzzy word features. We present a two-dimensional fuzzy word classification system where the spatial locations and shapes of the fuzzy membership functions are derived from the training words. The system is able to achieve an average recognition rate of 74% for the word being correctly classified in the top position, and an average of 96% for the word being correctly classified within the top five positions.

Finally, in Chapter 9 we summarize the handwriting recognition techniques presented in this book and discuss some possible and promising directions for further research. In particular, multi-level recognition systems, Markov random field models, and soft computing for handwriting recognition.

Handwriting recognition is not a simple signal processing task, rather, it is an intelligent process that requires learning, reasoning, and knowledge. If we were ever to develop a reliable and accurate recognition system, we cannot rely on traditional computational techniques. We need to use soft computing techniques for developing the new generation of handwriting recognition systems.

Contents

1

Introduction

Computer recognition of characters or word is one of the most successful applications in computer vision [180, 194, 201]. Put it simply, it is an automated process that uses pattern recognition and machine learning techniques to *recognize* characters or words given a lexicon or even an entire dictionary. With the ever increasing computational speed, memory, technological advances in scanning devices and maturity of recognition techniques, we have seen a tremendous progress in the development of handwriting recognition systems in recent years. Fueled by market demand, some handwriting systems are now available as commercial products. Despite all these, off-line handwriting recognition is still a major challenge in pattern recognition [248].

There are two major problem domains in handwriting recognition: on-line and off-line. In on-line handwriting recognition, data are collected while they are being generated on a digitizing surface. In addition to the spatial positions (x, y) indicated by the tip of the writing tool on the surface, the data also include velocity $v(t)$ and acceleration $\alpha(t)$ of the pen's (or stylus') movement. Based on $\{(x, y), v(t), \alpha(t), t\}$, the recognition algorithm then infers the written characters or words in *real time*. However, in off-line handwriting recognition, all that is available to the recognition system is the digitized spatial information, e.g., the image of the address scanned from an envelope or an amount shown on a cheque. As a consequence, on-line handwriting recognition has a much higher recognition rate as compared to that for the off-line case. On-line recognition techniques are now widely used in hand-held computing and communications devices, most notably, PDAs. Inspired by the fact that on-line recognition makes use of spatial-temporal features in the data, some researchers have considered extracting dynamic (temporal) information from off-line handwritten word [184, 134].

In terms of processing word data, there are two approaches [43]: The analytical approach which treats the word as an *orderly* collection of parts. In such approaches, the word image is segmented into subunits based on some properties or rules, e.g., character-like properties or shapes that match a part list (e.g., alphabet). Before recognition, the word has to be reconstructed, usually with the aid of a dictionary to correct misclassifications [17, 26, 138, 234]. Segmenting a word image into meaningful parts is itself a difficult task; there are no techniques that are able to reliably segment a word into characters. For example, a handwritten *u* may be segmented into alphabetically *meaningful* parts: *ii*, *il*, or even *ll*. Although extensive work has been done along this direction, it remains an open question as to whether it is the most promising approach [26, 85, 104, 152, 240, 244].

Another approach is the holistic approach which considers the word as a single entity [167]. This avoids the problem of character segmentation, but is an area which has been less studied than the analytical approach. However, research along this line has recently gained considerable interest [32, 92]. However, such schemes have typically used only binary images and assume that handwritten words have regular appearances. Despite the fact that, in a handwritten word, the formation of the characters is in most cases unconstrained. Humans recognize handwritten words based on the knowledge of possible character combinations. Therefore, the human reader is able to recognize the word with blurred or even missing characters (refer to Jacobs and Grainger for a recent review of the literature in the area of human models of visual word recognition [123]). Thus holistic approaches to handwriting recognition may be more natural and robust, and reliable if we take knowledge and inference rules into account. However, current holistic approaches are effective mainly for applications such as the reading of bank cheques, postal addresses, and forms, which require small or static lexicons, or used as a subsystem for lexicon reduction.

The performance of a recognizer is also very much dependent on the lexicon size: The larger the lexicon size the lower the recognition rate [99]. Since the recognizers developed so far are normally based on some similarity measures, bigger lexicon means there are more words sharing common (similar) features that may confuse the recognizer, resulting in lowered recognition rate. This also makes it difficult to make an objective evaluation of the recognizer.

Similar to any pattern recognition systems, the handwritten word recognition system contains the following basic components: pre-processing, feature extraction, modeling, and recognition. In pre-processing, the system carries out *data cleaning* tasks which may include noise reduction, normalization, skeletonization, and/or binarization, etc. Figure 1.1 shows a simplified handwritten word recognition system. This is usually followed by a segmentation process (not shown in Figure 1.1).

In this chapter we will present an overview of off-line handwriting recognition techniques.

1.1 Feature Extraction Methods

Feature extraction is of vital importance to handwriting recognition systems. It serves two main purposes: extracting the most representatives features that are used by classifiers, and reducing redundancies in data. In on-line systems, input is composed of unit width line-segments with dynamic information so that feature extraction is a relatively easy task. In off-line systems, feature extraction is affected by several factors [251, 235]:

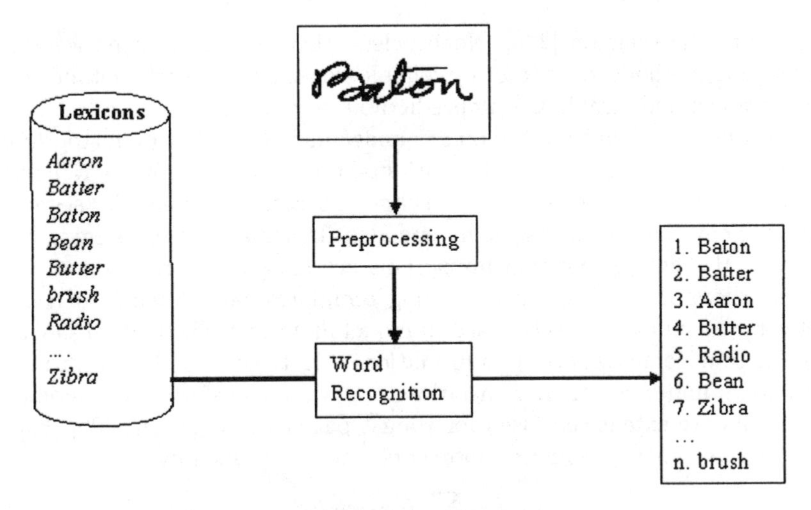

FIGURE 1.1. Handwritten word recognition system

- different backgrounds of documents;
- non-uniform illumination of the scanner;
- convolution distortion due to the point spread function of the scanner;
- noise introduced by electronics and writing tools;
- different qualities of paper and types of ink.

In order to overcome these problems, many methods have been proposed for off-line handwriting recognition.

1.1.1 *Extracting Features from Binary Images*

The majority of previous work has centered on using binarized images and skeletons or contours of the binarized image. There are a few good reasons to use binarized images: the scanning and processing equipment for binary images are cheaper, storing binary images requires less memory, and many good algorithms are available for extracting features from binary images. These approaches invariably lose information and are sensitive to noise and other degradations. When the binarized word is used, inner and outer contours are used to characterize the word structure [26, 234]. The threshold level needs to be continuously adjusted across the image to take account of changing intensity levels and to overcome feature omission or blurring problems. However, character borders are still subjective and open to interpretation. Furthermore, forming the skeleton of characters (or words) generally results in some important information being lost (where holes or cavities are filled, or parts are blurred), and can introduce distortions or spurs into the skeleton [236]. Nonetheless, there has been some work on developing methods for extracting reliable skeletons, but such methods can still produce undesirable and unpredictable results [28].

Features extracted from the binary image include projection histograms, zoning templates, and moment invariants, etc.. Projection histograms are used in recognizing segmented words and characters, and also for segmentation. They can be made scale independent by expanding or merging bins on axis. However, projection histograms are sensitive to rotation as well as variations in writing styles. Zoning templates can be fed into neural networks as input or directly used in model matching. The main disadvantage is that some important information is lost. Due to their invariance properties under rotation, translation, and scale transformation, moment invariants are extensively used for robust pattern recognition [206]. For a given binary image, geometric moments can be obtained by

$$m_{pq} = \sum_{M} f(x,y)x^p y^q, \tag{1.1}$$

where M is the image size, and $f(x,y) \in \{0,1\}$ is the image intensity function. The moment invariants were first proposed by Hu [115] and modified by Reiss [213].

Thinning or skeletonization is very popular as they can obtain thin-line representation of images for data compression. Thin-line representations are more amenable to extracting critical features such as end points, junction points, and connections among the components [117], whereas vector quantization algorithms often used in pattern recognition tasks also require thin-lines as input [150]. Usually, thinning algorithms can be divided into three categories: 1) distance transform; 2) iterative erosion or fire-front propagation from the boundary of images; and 3) analytical methods.

Distance transform-based thinning methods first obtain the symmetric axis transform of an image [253]. The symmetric axis of the image corresponds to the local maxima of the distance transform. Then the skeleton of the image can be obtained from the local maxima of the distance transform. These methods are computationally intensive, and require large memory space. Methods proposed by Arcelli [4] and Xia [255] based on iterative erosion or wave propagation from the boundary can guarantee the connection of the results, but may produce false skeletal branches. The methods in this category can be implemented using sequential or parallel schemes. Sequential algorithms examine the boundary points for deletion in a predetermined order and are able to preserve connectivity. For fully parallel thinning algorithms, connectivity must be carefully taken into consideration. Zhang and Wang developed a so-called *perfectly* parallel thinning algorithm that can ensure the connectivity of skeletons [260]. In their algorithm, a 3×3 window is used. Important parameters such as neighbor number $NN(p)$, weight number $WN(p)$, and connection number $CN(p)$ are defined. According to these parameters, all pixels in the image are classified into several point sets including end-point set, break-point set, hole-point set, removable-point set, and unremovable-point set, based on which, the algorithm can produce skeletons consisting of irreducible sets only. This algorithm is relatively faster but more complex in programming. Analytical methods are usually suitable for the skeletonization of simple shapes. For instance, thinning using Fourier descriptors can produce smooth skeletons but the shapes must be simple (without holes).

Contour representation of images is also popular for data compression. Contours of images can be easily obtained by tracing algorithms and many features can be extracted from contours efficiently. Geometric moments can be efficiently derived from chain-coded contours [129], because the amount of data for boundary representations is substantially smaller than that for region representations. Fourier descriptors can achieve translation, rotation and scale invariant pattern recognition for two-dimensional shapes [198, 148, 40]. The coefficients of Fourier descriptors can be easily extracted from closed contours.

1.1.2 Extracting Features from Gray-Scale Images

Binarization by any method is a lossy operation. When binarization is performed on gray-scale images by thresholding [218], especially at the early stage of the recognition process, it may lead to the loss of some significant information. Adaptive thresholding can be used to reduce information loss, in which the threshold level of binarization needs to be continuously adjusted across the images to adapt the changing intensity and illumination. However, no matter how good the binarization methods are, they often produce broken and touching line-segments.

Extracting feature directly from gray-scale images can overcome the disadvantages caused by binarization. Different algorithms have been designed to extract different features. Wang and Pavlidis proposed a method to extract features directly from gray-scale images [251]. They classify pixels into topographic structures (e.g., peak, ridge, saddle, flat regions) which are grouped together into regions. The interrelationships among these regions are extracted and analyzed, and polyline approximations are then used to form stroke skeletons. Pettier and Camillerapp also used gray-scale images to form skeletons by placing the skeleton between the edge points located at the Laplacian zero crossings, which is then analyzed to provide good connectivity [202]. They found that with gray-scale images they could further process segmentation defects, which cannot be done with binary images.

Their method includes three steps. The first step is to assign each pixel to one of the topographic features according to the first and second derivatives and to the eigenvalues of the Hessian. In the second step, a topographic feature graph (TFG) is established in a single pass. Each node of the TFG is connected with a *characteristic* region that can be a ridge region, saddle region, or a flat region. In the last step of feature extraction, the skeleton of the image is formed through grouping and assembling the points and line-segments obtained from TFG. Lee and Kim [156] proposed a method based on the same idea presented in [251]. Their method can extract topographic features directly from gray-scale images. They claimed that their method improves the performance of feature extraction as compared to that of Wang and Pavlidis' method. However, the first and second directional derivatives at pixel P in these methods are computed based only on two or three boundary pixels, respectively. As a consequence, this method is susceptible to noise, especially in flat regions. Further, these methods cannot be applied to binary images, because the first derivative of the intensity at P is equal to zero if P is not an edge pixel.

Many features widely used in binary images can also be used in gray-scale images with minor modifications. For instance, geometric moments can be obtained by (1.1), where the range of $f(x, y)$ is between 0 and 255. Geometric moments are also invariant to contrast, but they are not illumination-invariant.

1.2 Pattern Recognition Methods

In general, there are three basic approaches: statistical pattern recognition, structural pattern recognition, and soft computing for pattern recognition.

1.2.1 Statistical Pattern Recognition

In the statistical (or decision-theoretic) approach, the recognition is based on the decision boundaries established in the feature space by statistical distributions of the patterns [209]. A decision is usually made by maximizing *a posteriori* probability, where the recognition error of this approach is called Bayes error. In this approach, the observation can be represented as a set of random variables and the density function of the observation depends on its class. If the observation consists of a certain number of features, the classifier is usually based on minimizing the overall Bayes error. It is well known that finding the global minimum point is not an easy task as the surface of the Bayes error is complex. However, if the observation can be expressed as a set of sequential features or vectors, some techniques, such as hidden Markov models, can be used to exploit the sequential property to reduce classification errors.

The theory of hidden Markov models (HMMs) was first developed in the 1960s by Baum and his colleagues. As their papers [10, 12] were published in the journals on mathematics, HMMs remained largely unnoticed by the pattern recognition community until the mid 70s. Thanks to the pioneering work of Baker and Jelinek [8, 127] and the inspiring tutorial papers of Rabiner and Picone [209, 199], HMMs have been widely used in speech recognition. In recent years, HMMs have been applied to handwriting recognition. The early attempts at using HMMs to recognize handwritten words were made mainly at the word-level for on-line applications [82], because the dynamic information about the writing process is available. The movement of the pen with respect to the tablet can be represented by a time function that can be sampled at a certain rate to form a sequence of features. HMMs are well suited for modeling such sequential features. This makes it possible to apply HMMs to on-line handwriting recognition [168]. However, if a system is based on word-level HMMs, it needs at least one model for each class. For a given class or word, the system needs a considerable amount of samples for training, otherwise the performance of the system may fall dramatically. Usually, systems based on word-level models are appropriate for tasks with small lexicons.

Recently, many systems use letter (character) models instead of using word models [114, 182]. The main merit of character-level models is that a limited number of models can recognize words with a very large lexicon. Word recognition is performed by matching a possible sequence of character models with the help of a given dictionary. If a word is segmented into characters or sub-characters before the recognition stage, dynamic programming can be used as the matching algorithm [230]. If no segmentation is involved, a level-building algorithm can be used to recognize words [181, 210]. As level-building algorithms segment words based on maximizing global output probability, they are able to achieve a better performance. For Chinese character recognition, a similar principle can be applied. There are about 55,000 characters in Ci-Hai dictionary and about 5,000 characters used daily in China, but there are relatively fewer structural parts called *radicals*. Therefore, it is believed that the complexity of Chinese characters can be simplified if processing and recognition are done according to the radicals [54]. Also, the number of models can be largely reduced. Kim and others developed a system based on HMMs of radicals and level-building algorithm to recognize Chinese characters in Korea, the correct recognition rate was 90.3% on their testing database [137]. A similar system was also developed to recognize Korean characters [229]. The main weakness of these systems is that they are sensitive to the writing order of strokes.

HMMs are also widely used in off-line handwriting recognition. Projection histograms are popular for transforming two-dimensional images into one-dimensional feature spaces. Park and Lee proposed to use four projection histograms obtained by different transformations for one image, and then use four HMMs to model these projection histograms [190]. In their system, the final decision is made based on the outputs of HMMs by three methods: unequal-weight combining, equal-weight combining, and majority voting. They found that the recognition rate of the best single classifier is low (61.3%); however, with the combination of the classfiers, the system's performance can be as high as 91.6% in recognition accuracy. However, projection histograms are sensitive to rotation and slant, and different characters may produce similar projection histograms as some important information is lost during feature extraction.

Other popular methods are based on the assumption that a character image can be represented by the profile generated by observing a sequence of horizontal or vertical scan lines through the image [80, 81, 143, 60]. The ordering of features can also be done by some heuristic rules [48] or around the contours [36]. Cho, Lee and Kim [60] proposed to split a word image into a sequence of thin vertical frames that are encoded into a sequence of discrete symbol using vector quantization [169]. They used a forward-backward method to search for the best path. Because the standard HMMs can not guarantee to produce a legal sequence of characters, ligature models or word models based on spelling are commonly used to constrain the search space [60, 153]. Elms used two HMMs to represent the features extracted from columns and rows, respectively [80]. The decision was made based on the product of the two models' output probabilities. Elms' method achieved a good performance that was evaluated on a machine printed word database and a handwritten numeral database. Elms found that the combination of HMMs outperformed either of the two models.

Kuo and Agazzi developed new methods based on pseudo 2-D HMMs to recognize printed documents [149]. A pseudo 2-D hidden Markov model can be viewed as a macro-HMM, and its states are represented by micro-HMMs. In their system, features were extracted by zoning. Experimental results on machine printed words showed that the system based on the pseudo 2-D HMMs achieved higher recognition rates than that based on the standard HMMs. State duration can be used to further improve the performance of handwriting recognition systems due to the variations in character width. The state duration modeling was first proposed in speech recognition by Levinson [157] and then applied to handwriting recognition by Chen, Kundu and Srihari [48]. Vaseghi proposed another method to model state duration [247]. Vaseghi's method requires more computer memory, but it does not increase much computation. This method was adopted in handwritten numeral recognition with good results [36].

If a system aims at recognizing words within a certain context, language processing is useful. Usually language processing is constrained by application domains, e.g., speech recognition, on-line or off-line handwriting recognition. In conventional systems, language processing is employed as a post-processing of the recognizer, therefore the recognition process is not optimal in the usual sense. With the development of stochastic language modeling techniques [165], the language process and the matching process have been merged into one process. In large vocabulary word recognition tasks, the system should be able to search N-best paths of HMMs, because the N-best paradigm can achieve a significant reduction of the search space [221].

1.2.2 Structural Pattern Recognition

In structural (syntactic) approaches, each pattern class is defined by using structural descriptions or representations [32, 89]. The recognition is performed according to structural similarities. This is based on the fact that structural relationships between features are also essential to recognizing patterns. Graphs offer a general framework for encoding pictorial data. This has motivated a large amount of work in graph grammars and languages [195].

An important early work for character recognition is the analysis-by-synthesis method proposed by Eden [78]. In his method, all Latin script characters can be formed by 18 strokes and generated by grammar-like rules. The recognition is based on a generative process. Pavlidis and Ali utilized the polygonal boundary approximation for shape recognition and applied their method to handwritten character recognition [197]. In this method, classifiers recognize characters by analyzing the extracted features. Therefore, this method relies heavily on semantics. Fu and Booth gave a good survey of the early work on grammatical inference for syntactic pattern recognition [87, 88]. In real-world applications, a system should be able to automatically learn grammar rules or extract structural relationships from training samples. However, the algorithms for extracting grammars or graphical descriptions are not well developed. A simple and widely adopted remedy is to use fuzzy features, stochastic grammars [89] or fuzzy relational graphs [32].

Recently, there are a considerable number of systems using fuzzy grammars and graph trees for handwriting recognition. Abuhaiba and Ahmed proposed to use a set of 105 fuzzy constrained character graph models, which can tolerate large variations in writing styles, to recognize unconstrained off-line handwritten numerals [1]. In their method, a polygonal approximation of a character image is obtained from its skeleton. Then the polygon is converted into a tree structure (graph). A node of the tree structure can be a line segment, an intersection, a single loop or two adjacent loops. The features of a node are represented by fuzzy sets. Their method was evaluated on a USPS database and achieved an overall correct recognition rate of 90.7% with a rejection rate of 6.4%. Abuhaiba, Datta and Holt modified the method in [1] to recognize unconstrained handwritten strokes [2]. In this method, they used fuzzy features to describe two-dimensional shapes and used states to replace the tree nodes. As a result, the performance of the system was improved. The recognition and rejection rates were 94.8% and 1.2% respectively. Malaviya and Peters developed a fuzzy language (FOHDEL) for on-line handwriting recognition [170]. They used three fuzzy sets of features including geometric features, positional features and global features to describe handwritten patterns. The classifier makes the decision by a feature aggregation process according to some fuzzy linguistic rules. On the Latin lowercase alphabet test set and the

Arabic numeral test set, they achieved a recognition rates of 93.7% and 95.1%, respectively. Parizeau, Plamondon and Lorette presented a method for cursive script segmentation and recognition based on intrinsic models of cursive letters (allographs) [193]. The models were built using stratified context-free shape grammars that permit the definition of both syntactic and semantic attributes. These attributes synthesize pertinent morphological characteristics of allographs. In the parsing process, they used fuzzy consistency and adjacency constraints. In their experiments, they developed a set of intrinsic letter models for all 26 lowercase alphabetical letters, which were optimized with a multi-writer isolated letter database. They reported an average recognition rate of 91.7%.

Matching graphic structures and relationships between components without parsing is also a popular approach in handwriting recognition. Rocha and Pavlidis developed a character recognition system based on graph transformations [215]. Their system extracts skeletons directly from gray-scale images to form graphs. To compute the graph matching cost, they took the conceptual modeling of shape variations, distortions, noise, etc. into account. The overall cost is the summation of several types of the weighted transformation costs. In their system, they took a great care in designing graph transformations, giving the filled gaps, broken lines and distortions a careful consideration. Encouraging results were obtained in [215]. However, the system was not able to generate prototypes in a fully automatic way. Nishida and Mori proposed an automatic algorithm for character shape prototype construction [183]. They decomposed images into singular points and primitives from which they constructed primitive sequences (PS). The PSs are then labeled for further process. The structural models of a class was defined on the basis of the relations in the PS-label set and singular point structure set. They defined similarity measures by algebraic representations of continuous shape transformation based on structural models. A recognition rate of 95.86% was obtained on a database of handwritten numerals.

Recently, structural classifiers have also been used in handwritten word recognition. Buse, Liu and Caelli developed a handwritten word recognition system on the basis of the similarities between relational graphs of word images and corresponding prototypes [32]. In this system, they represented the structure of a word in terms of the relational graph that encodes a set of oriented segments in the word and the relationships between the oriented segments. With only a few training samples per class, their method obtained good results.

Structural matching plays a major role and is widely used in Chinese character recognition. Chou and Tsai presented a method for recognizing handwritten Chinese characters by matching stroke segments [64]. The stroke-segment similarity was defined on the likelihood between the features of any two stroke-segments. They used an iteration scheme to measure the overall similarity between the character and the template, which adjusts the matching relationships between the input and template stroke-segments using the contextual information. The final matching relationships were determined by a mutually-best match strategy and the final similarity measure was obtained accordingly. They conducted experiments on a Chinese character database of 465-lexicon size and achieved a recognition rate 96%. Cheng proposed to use multi-stroke relaxation matching for handwritten Chinese character recognition [56]. In his method, the similarity between two strokes was measured based on affine transformation. The matching algorithm is divided into two processes: the first process computes the matching probability between two strokes, including stroke merging; and the second selects the optimal matching pairs according to the probability matrix. Cheng tested his system on a large lexicon of 2000 handwritten Chinese characters and reported a recognition rate of 93.8%.

1.2.3 Neural Networks for Pattern Recognition

Since the mid 1980s, neural networks have become popular in handwriting recognition. Among them, the feed-forward neural networks remain dominant because the well-known training methods, back propagation (BP) of errors and function approximation, are available.

Multi-layer perception (MLP) classifiers are able to form complex hyperplane decision regions. Most MLP neural networks have only one hidden layer [151]. Inputs to neural networks may vary from system to system. They can be raw images, skeletons of digits and other features. LeCun and his colleagues used an MLP classifier with three layers where the hidden layer and the output layer were fully connected [151]. The recognition rate for 2,007 handwritten numerals was about 94%. In their second experiment, they used an MLP classifier with five layers (three hidden layers) to recognize handwritten numerals. The first two layers perform two-dimensional convolutions over the image. Therefore this neural network was able to extract local features and combine them to form higher-order features. They were able to achieve a recognition rate of 95%. Lee developed a system based on multi-layer cluster neural network [154]. They extracted oriented components from binary images using the Kirsch masks [205]. The input layer and hidden layer were locally connected and the hidden layer and output layer were fully connected. The error back propagation training algorithm [262] was used to estimate the weights of neural nodes. The initial weights were determined by the genetic algorithm [15]. They tested their system on the handwritten numeral database from Concordia University in Canada and achieved a recognition rate of 97.10%.

Radical basis function neural networks use localized basis functions (kernel functions) as their weight functions. The basis functions form complex decision regions from overlapping receptive fields created by kernel function neurons. The output of a neuron (node) in the hidden layer has a peak when the input vector is similar to the center of the basis function and falls off monotonically as the distance between the input vector and the center of the basis function increases. Typically, radical basis function (RBF) neural networks have one hidden layer of kernel function nodes and have a bias input for the output layer. The function of output nodes is typically the sigmoid function or a summation function. For RBF neural networks (RBFNNs), the kernel functions can be biased or unbiased exponential functions. In [20], RBFNNs with both types of kernel functions were evaluated under different experimental conditions. They found that RBFNNs with biased kernel functions obtained better results than those with unbiased kernel functions under optimal conditions. It was reported [120] that training algorithms for RBF neural networks are much faster than that for MLP neural networks, whereas the correct recognition rates of the two types of neural networks are similar.

Probabilistic neural networks (PNNs) are relatively new compared to the MLP and RBF neural networks, although the topologies of these networks are similar. The essence of the PNN is the Bayesian paradigm. In PNN, the activation function in pattern units is an exponential function instead of the sigmoid function as in the MLP. The PNN performs classification by finding the maximum *a posterior* probability (MAP). The performance of the PNN is determined by its decision regions that are closely related to the accuracy of the estimation of probability density functions. Usually, the probability density functions are assumed to be Gaussian functions or Gaussian mixtures that can be used to approximate any probability density functions. Typically, PNN uses two or three hidden layers to obtain the accurate estimation of the true probability density functions. The major advantages of the PNN are that its training process is very fast and it can approach the Bayesian optimum on the decision surface. As a result, the PNN is able to cope with variations in data samples. Due to these merits, the National Institute of Standards and Technology performed studies with PNN and achieved high performance in handwritten digit recognition [231]. A performance comparison of the three neural classifiers on a handwritten numeral database was given in [20]. A conclusion was drawn from the comparison that the probabilistic neural network outperforms MLP and RBF neural networks in both speed and recognition rate.

1.2.4 Soft Computing in Handwriting Pattern Recognition

In comparison with the three approaches introduced in the previous sections, there is a strong evidence that structural approaches and neural network-based approaches may offer a better solution to the problems in handwriting recognition. Statistical approaches are more suitable to the problems with large random variations in data, such as speech recognition, because these approaches have their own limitations; for instance, HMM is one-dimensional in nature and has difficulties in modeling structural information, whereas words and characters are highly structured entities. In addition, many uncertainties in handwriting are not random by nature or may be difficult to model by traditional probabilistic techniques. For such uncertainties statistical approaches may fail. These traditional approaches are based on *crisp* (hard, binary) computational paradigms that cannot effectively handle many types of uncertainty and deal with partial truth that occur frequently in nature [261]. In order to develop handwriting systems that are able to make use of statistical and structural information and can handle different types of uncertainty, we need to combine different paradigms. Recently some researchers have used soft computing approaches to developing hybrid handwriting recognition systems.

Soft computing is a new problem-solving paradigm that combines emerging techniques and theories such as fuzzy logic, neural networks, genetic algorithms, evolutionary computation, and probabilistic methods for complex problems that cannot be solved by conventional mathematical methods [22]. Such handwriting recognition systems can be classified into two categories: In the first category, they recognize patterns based on a single classifier that is formed by combining different soft computing methods which may include neural networks, fuzzy logic, genetic algorithms, and probabilistic modeling. The recognition systems belonging to the second category make decisions based on results of several classifiers.

In the first category, the systems usually use different methods to compensate each other. Chiang and Gader used two three-stage hybrid systems to recognize handwritten words [52]. In both systems, the first stage is to use a self-organizing neural network to convert multi-dimensional bar feature vectors into two-dimensional feature maps (SOFM) [147]. Thus, the features of a whole character yield an allograph map. At the second stage, the allograph map is converted into two allograph fuzzy membership maps, each of which corresponds to one of the two systems. At the third stage, the allograph fuzzy membership map is used as the input to a fully connected MLP neural network in the first system. For the second system, a fuzzy membership of the character is computed by using a bank of Choquet fuzzy integrals. With soft computing techniques and soft segmentation, they were able to achieve better results as compared to the more traditional approaches.

In [252], a hybrid neural-HMM system was proposed to recognize off-line cursive words. In their system, feature sequences are extracted from vertical frames or slices of images from left to right. The neural network uses the three previous frames to predict the feature vector of the current frame. The sequences of prediction errors are used as the input to the pseudo HMMs. The decision is made by the HMMs that minimize the overall prediction error instead of maximizing the *posteriori* probability. Because some structural information lies in the prediction errors that can be modeled by HMMs, it is reasonable to obtain a better performance by combining neural networks and HMMs. Cho also developed a system based on HMMs and MLP neural networks [59]. In conventional HMM-based systems, the recognition decision is made by finding the maximum of the output *posteriori* probabilities. However, in the HMM/MLP system, all output *posteriori* probabilities of HMMs are used as the input to the MLP neural networks [59]. Therefore, the system is able to utilize mutual information to enhance its performance, but it is difficult for HMM-based systems to do so. Cho compared four classifiers listed in the order of recognition rates from the highest to the lowest: HMM/MLP, SOFM, HMM and MLP.

In the second category, a system combines the decisions of several classifiers to produce higher recognition rates. Hashem and Schmeiser proposed to use optimal linear combinations of corresponding outputs of neural networks [107]. The weights were derived by minimizing the mean squared error with respect to the distribution of random inputs. However, they did not conduct any experiments to prove the effectiveness of their system. Cho and Kim presented a method using fuzzy logic to combine the results of several neural network classifiers [58]. In their method, the fuzzy integral is a nonlinear function that is defined with respect to a fuzzy measure based on the performance of several classifiers. Their experiments on handwriting recognition showed that with the fuzzy integral their system's performance was improved considerably.

In the above methods, the multiple classifiers used for combination must belong to the same type of classifiers (e.g., neural network based classifiers). Therefore, it is difficult to apply these methods to combining different types of classifiers. In order to deal with this problem, another scheme was proposed in [110], in which the decisions were made according to the rankings of a given class set. The combination methods were either to reduce or re-rank the class set. Because this scheme allows the combination of different types of classifiers, it is possible to use different features for different classifiers so that the classifiers can be designed to complement each other. In [110], several systems using different combinations of classifiers were tested on machine-printed character and word and handwritten numeral databases. Their experiments showed that all the combinations outperformed any individual classifier in the combinations.

Kittler *et al.* developed a common theoretical framework for combining classifiers [145, 143]. They developed four individual classifiers: structural, Gaussian, MLP neural network and HMM classifiers and carefully examined six combining rules: majority vote, sum, max, min, product and median. In their experiments, they found that the simple *sum rule* and *median rule* obtained the best performance and the *min rule* and *product rule* resulted in the worst experimental results. Rahman and Fairhurst evaluated several combining schemes for handwriting recognition [211]. In the parallel combining scheme, individual classifiers not only hand their decisions about the identity of the numeral to the decision fusion layer but also send their rejected decisions to the re-evaluation part of the system for further examination. The system reached the final decision by combining both accepted and rejected decisions. The experiments showed that the combination of classifiers increased recognition rates to a certain degree. The combination of classifiers based on rules definitely improves system performance. However, the design of rules requires human knowledge and a careful analysis of classifiers.

Freund and Schapire proposed a promising boosting algorithm, AdaBoost, that can be used to automatically construct a composite classifier by sequentially training classifiers [84]. AdaBoost combines a series of learning machines (weak learners [84]) each having a relatively poor recognition rate to form an ensemble (a strong learner [84]) that has a good performance. In conventional methods, the error rate decreases as the number of weak learners increases on the training data. However, error rates on the test data will keep increasing if the number of weak learners exceeds a certain threshold. AdaBoost can overcome this problem, as it is insensitive to the number of weak learners. In [73], ten ensembles were used to recognize handwritten numerals. Each ensemble was used to separate a particular digit from other digits. Every ensemble consists of a number of boosted trees. The performance of the ensemble for a particular digit is much better than that of single trees. Consequently, the overall recognition rate is very high [73]. Schwenk and Bengio explored the application of AdaBoost to auto-associative and MLP neural networks on handwriting recognition [222]. AdaBoost combines the results of classifiers trained consecutively. Each classifier is trained with an emphasis on some particular classes according to a cost function that is weighted by margin distributions. An important property of such a system is that the bound on the generalization error is relatively independent of the number of combined classifiers. Experimental results showed that its recognition rates initially increased then became steady with the further increase of the number of classifiers on both training and test data, which is consistent with the theoretical property of AdaBoost. Compared with the MLP neural networks, the error rate on the on-line test data has been reduced from 6.2% to 2.0%. From the bound of the generalization error, we can find that the increase of the margin is equivalent to the reduction in classifier complexity. There is a balance between the training error and complexity [174]. Based on this discovery, Mason, Bartlett and Baxter developed a "DOOM" algorithm [174] that can achieve better results than the standard AdaBoost algorithm.

2

Pre-processing and Feature Extraction

Many different types of features can be used to recognize handwritten words and characters. Good features should enable the system to discriminate different classes effectively, to reduce redundancy in representation and be robust to noise and deformation. In this chapter we discuss features and feature extraction techniques for handwriting recognition.

2.1 Pre-processing of Handwritten Images

Due to large variations in handwritten image data, pre-processing, which is designed to reduce noise and variations in the data and produce a more consistent set of data for recognition, is necessary for accurate handwriting recognition.

2.1.1 Pre-processing for Handwritten Characters

In the discussion of pre-processing techniques for handwritten characters we will be more concerned with methods for connecting disjoint regions, slant/tilt correction, and size normalization, because they have significant effects on the success of the recognition system.

Disjoint Region Connection

Broken lines degrade the performance of handwriting recognizers [139]. In HMM-based handwritten numeral systems [37, 39], features are extracted from outer contours of numerals. An ideal outer contour is a closed curve so that the extracted features can be arranged in a sequence around the curve, which is more amenable to modeling. Many methods for connecting disjoint regions are available in the literature. Broadly, the connection process can be summarized as follows:

1. Find the disjoint regions, label the biggest region, and detect prominent points of curves, where the prominent points are the local points that have maximum curvatures.

2. Calculate the distance between the prominent points in the biggest region and other regions.

3. Find the *closest* pair of prominent points (p_0, p_1), where p_0 is in the biggest region and p_1 is in the smaller region.

4. Move the smaller region to the biggest one in the direction from p_1 to p_0, so that they can be merged into one region.

5. Go to step 3 if there still exist disjoint regions.

After this process, there will be only one outer contour.

Slant Correction for Characters

A practical character recognizer must be able to maintain high performance regardless of the position, size and slant of a given character or word. For handwritten characters, a major variation in writing style is caused by slant that is defined as the slope of the general writing trend with respect to the vertical line. It is important that the recognition system be insensitive to slant. Broadly, there are two ways to achieve this: (1) using invariant features; (2) removing variations, which produces a more consistent set of features and thus improves the performance of the recognizer.

It is generally difficult to find features that are *truly* invariant to slant. Rotation invariants may be used to approximate slant invariants only if the slant is small enough. Moreover rotation invariants cannot be used to distinguish some characters, e.g., "6" and "9". For variation removal, some systems use second-order moments as parameters to estimate the rotation from isolated characters [16]. The estimated rotation is given by

$$\angle rotation = \frac{1}{2} \arctan 2M_{11}/(M_{20} - M_{02})], \qquad (2.1)$$

where

$$M_{pq} = \sum_{0}^{W-1} \sum_{0}^{H-1} (x - \overline{x})^p (y - \overline{y})^q f(x, y),$$

$(\overline{x}, \overline{y})$ is the centroid of the image, $f(x, y)$ is the gray value of the pixel (x, y), H and W are respectively the height and the width of the character image. This method works in some cases. Figure 2.1(b) shows such an example. Unfortunately the character's height is not always much greater than its width, and usually the rotation is not small enough. Methods based on rotation correction may fail in such cases as shown in Figure 2.1(e). The method based on eigenvectors of images also suffers from this problem, as these methods are designed for correcting rotations [80]. Cai and Liu proposed a method for slant estimation using linear regression [37]. They first calculate the centroid (x_{0j}, j) for every row of a given image and obtain H row centroid points, where H is the height of the image. Then, linear regression is used to best fit a line, $x = a + by$, by minimizing the sum of the weighted distances from the centroid points of rows to the line. The estimated slant is the slope of the line, which is given by

$$\angle slant = \text{ctan}^{-1}[S_{xy}/S_{yy}], \tag{2.2}$$

where

$$S_{yy} = \sum_{j=0}^{H-1} G_j(j-\overline{y})^2,$$

$$S_{xy} = \sum_{j=0}^{H-1} G_j(x_{0j}-\overline{x})(j-\overline{y}),$$

$$x_{0j} = \sum_{i=0}^{W-1} f(i,j)i / \sum_{i=0}^{W-1} f(i,j),$$

$$G_j = \sum_{i=0}^{W-1} f(i,j)\eta(|j-\overline{y}|),$$

and $\eta(\cdot)$ is the weight. The slant-corrected word image can be obtained by the following transformation:

$$x' = x - y/\tan(\angle slant), \qquad y' = y.$$

This method produces consistent slant-corrected images that are crucial to the performance of handwriting recognition. Figures 2.1(c) and (f) show that this method works very well as compared to those shown in Figures 2.1(b) and (e). This is due to the fact that this method considers slant instead of rotation as the main source of distortion, which occurs in most handwriting styles.

Size Normalization

Size normalization is used to reduce size variation. Directly scaling all images to an identical size will result in significant deformation in many cases [196]. According to observations, the character's height is usually greater than its width. Therefore a heuristic rule is set for scaling: If $W/H < 0.8$, the scales in horizontal and vertical directions are the same, otherwise the scale factor is set to $W/H = 0.8$. The mapping function for size normalization must adapt to scales. If we were to use the following simple mapping function:

$$f_s(x,y) = f(x/s_x, y/s_y), \tag{2.3}$$

where x and y are coordinates of images after mapping and s_x and s_y are scales in horizontal and vertical directions, respectively, we would obtain a

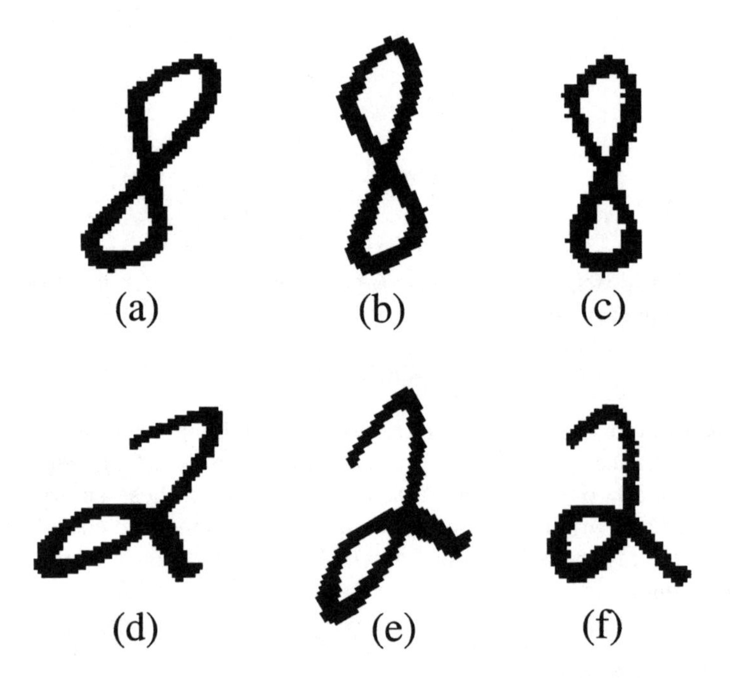

FIGURE 2.1. Slant correction. (a) and (d) the original characters; (b) and (e) the results based on rotation correction; (c) and (f) the results based on slant correction.

mapped character image that would be fragmented. Cai and Liu proposed a new mapping technique as illustrated in Figure 2.2(b) [37]. For instance, if $s_x \geq 1$ and $s_y \geq 1$, (2.3) is used as the mapping function. If $s_x < 1$ and $s_y < 1$, the following equation is used:

$$f_s(int(s_x x'), int(s_y y')) = \max_{x'} \max_{y'} f(x', y'), \tag{2.4}$$

where x' and y' are coordinates of images before mapping, $int(\cdot)$ returns the integer part of its argument: $x = int(s_x x')$ and $y = int(s_y y')$. Figure 2.3 shows an example of the pre-processing results.

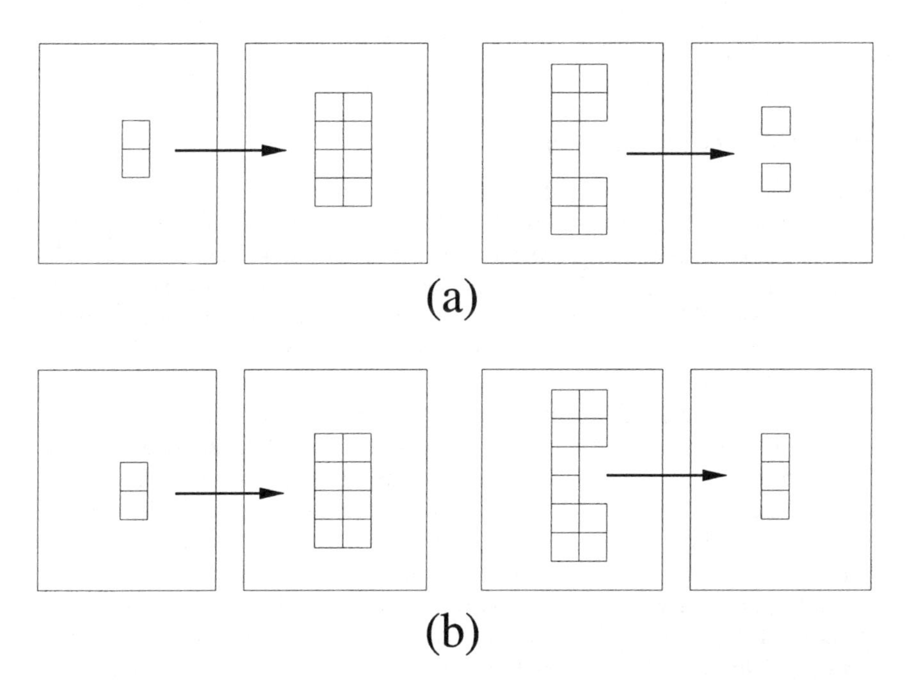

FIGURE 2.2. The mapping technique for size normalization. (a) scale-up and scale-down in conventional method. (b) scale-up and scale-down in proposed method [37].

2.1.2 *Pre-processing for Handwritten Words*

The pre-processing for handwritten words consists of slant correction, tilt correction, baseline finding, image normalization, and line-width calculation.

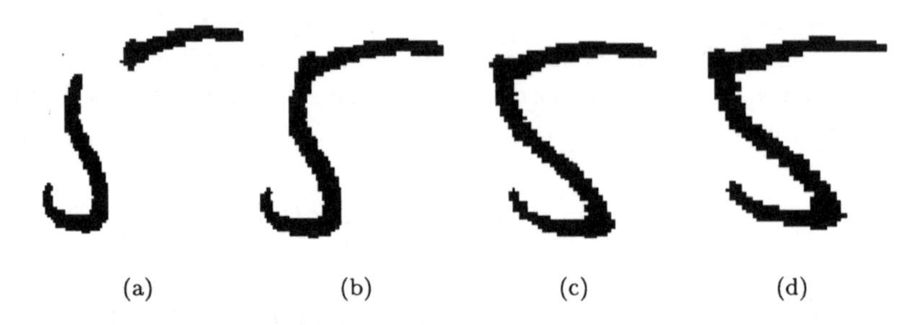

FIGURE 2.3. The pre-processing results. (a) original image. (b) image after broken line connection. (c) slant-corrected image. (d) size normalized image.

Slant Correction for Words

Similar to characters, word slant is defined as the average slope of a word with respect to the vertical direction. Before feature extraction, we must remove slant in the written word as much as possible [32, 26]. This will make recognition more effective, because we are able to extract a more consistent set of features. In addition, slant correction significantly reduces the difficulties encountered in character segmentation that is necessary for segmentation-based techniques.

There are several well-known slant correction algorithms in the literature. Bozinovic and Srihari estimate the word slant from binarized images [26]. They remove all horizontal lines by horizontal strip bars and discard all short lines. These strip bars also divide the image into parts. The slant of the word is defined as the average slant of all parts. This algorithm may fail, however, when a word has an obvious ascender or descender. Recently, Buse, Liu and Caelli proposed a more robust algorithm to determine word slant in the frequency domain [32]. This algorithm (to be presented in detail in Chapter 7) first uses the Fourier transform to convert an image into the frequency domain, then uses bilinear interpolation to change the Cartisian coordinate system to the polar coordinate system. Because the global orientation of the word is exhibited in its spectrum, the slant of the word can be calculated from the angular histogram of the spectral magnitude image.

Based on the idea of horizontal line removal in [26], Cai and Liu proposed a new efficient and robust algorithm to determine the global slant of a word, which is able to remove the interference by horizontal lines even for words with considerable ascender or descender [38]. For a given image, they first binarize the gray-level image using an adaptive threshold method [187], they then use a four-neighbor mask as shown in Figure 2.4(a) to detect edges. For a nonzero point, its neighbor number in the 4-neighborhood system is defined as follows:

$$N_4(p) = \sum_{i=1}^{4} n_i. \tag{2.5}$$

If $N_4(p) < 4$, this nonzero point is an edge point. Figure 2.5 shows an example for edge detection.

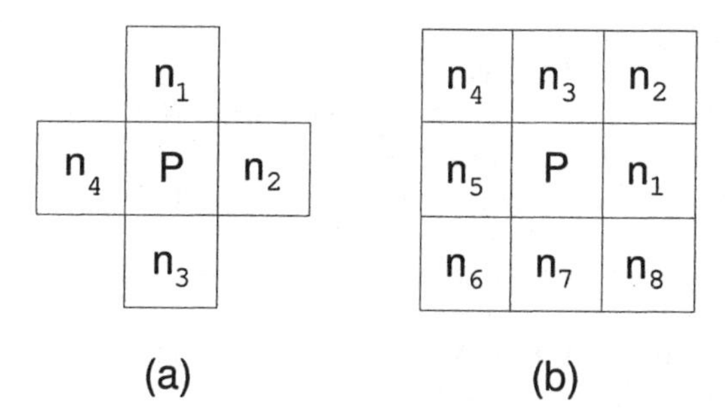

FIGURE 2.4. (a) 4-neighborhood of point P; (b) 8-neighborhood of point P.

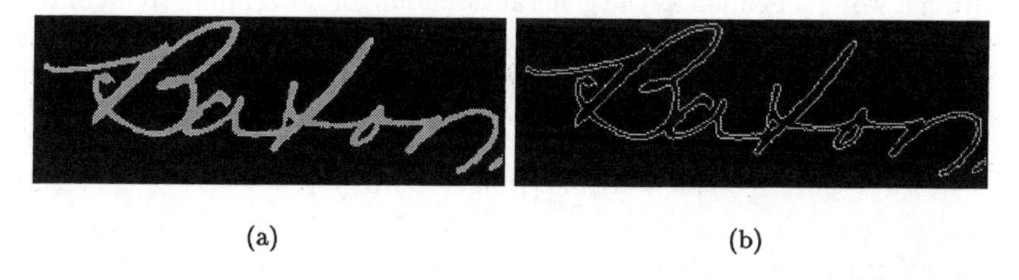

(a) (b)

FIGURE 2.5. Examples of edge detection. (a) binarized image; (b) edge of the binarized image.

However, the edges cannot be directly used for slant estimation, because they contain cross points and prominent corners where the orientation of the edge changes dramatically. The cross points are detected by a 3×3 mask shown in Figure 2.4(b). The neighbor number in the 8-neighborhood is defined as follows:

$$N_8(p) = \sum_{i=1}^{8} n_i. \tag{2.6}$$

If $N_8(p) > 2$, the nonzero point is a cross point. Corner detection is performed using the method proposed by Cheng and Hsu [55]. After deleting cross and corner points, an edge curve is represented by line segments. The endpoints of a line segment are located by a heuristic condition, $N_8(p) = 1$.

The length and the angle of the line segment are determined by its endpoints. Now, the global slant of a word can be calculated by

$$slant = \frac{\sum_{\theta_i \in S} l_i \theta_i}{\sum_{\theta_i \in S} l_i}, \tag{2.7}$$

where l_i is the segmented line length, θ_i is the angle between the segmented line and the x-axis, and the support S is in the range [30°, 150°]. The designed support range excludes all horizontal lines. This algorithm is simpler and more reliable than that proposed in [26]. The slant-corrected word image can be obtained by interpolation and the following transformation:

$$x' = x - y \times \arctan(slant), \quad y' = y, \tag{2.8}$$

where x', y', x and y are the Cartesian coordinates. Some results of slant correction are shown in Figure 2.6 and Figure 2.7.

Tilt Correction

Tilt of a word is defined as the general ascending or descending trend of the writing with respect to the horizontal line or the x-axis. Similar to slant correction, tilt correction also helps reduce variations in writing styles. Tilt estimation is based on the method in [32]:

$$tilt = \arg\max_{\alpha \in T}\{\max_{j \in Y} FD_s(\alpha, j)\}, \tag{2.9}$$

where T is the set of angle values, Y is the range of vertical coordinate, and $FD_s(\alpha, j)$ is the first derivative of smoothed projected-density histogram of the slant-corrected image at the angle of α and the position of $y = j$. The range of support Y is selected from the bottom to the centroid of the image. In general, tilts in most handwritten words do not exceed 10°, although the maximum word tilt angle in the database of [38] is 13°. Cai and Liu set the allowable tilt range, $T \in [-15°, 15°]$. Tilt correction is accomplished by

$$x'' = x', \quad y'' = y' - x' \times \tan(tilt). \tag{2.10}$$

However, the horizontal lines strongly interfere with tilt estimation, producing undesirable peaks in smoothed projected-density histograms. In order to avoid such interferences, Cai and Liu removed all of the horizontal lines by using Gabor filters, which will be discussed later. The horizontal line removal criterion is defined as follows:

$$p(x,y) = \begin{cases} 0, & \text{if } E^{0°}(x,y) \geq 2E^{90°}(x,y), \\ p(x,y), & \text{else,} \end{cases} \tag{2.11}$$

where $p(x,y)$ is the gray value of image at (x,y), $E^{0^\circ}(x,y)$ and $E^{90^\circ}(x,y)$ are the output energies of the Gabor filters oriented at 0° and 90° and at the position (x,y). For comparison, Figure 2.6 shows the results of tilt detection and correction with or without horizontal line removal.

In this example, there is an 8° difference between the two estimated tilts of the same word. The procedure described in (2.9), which is to the locate maximum peak of the first derivative of the smoothed projected-density function, though simple and effective, the horizontal lines in words may lead to significant errors. Indeed, removing horizontal lines helps improve tilt correction. Figure 2.7 shows more examples of the results of slant and tilt correction algorithms.

Baseline Finding

It is well known that spatial positions of word features are very important for word recognition, because this information can greatly narrow the choice of words [26]. In this chapter, spatial positions are defined with respect to four baselines: top, upper, lower, and bottom, as illustrated in Figure 2.8.

Bozinovic and Srihari proposed to use thresholds to determine the shoulders of the density histogram [26]. However, there are two major problems with their algorithm. First, the k highest density values have to be discarded. Different words and writing styles produce different number of peaks in the density histogram. Discarding too few or too many highest density values will result in wrong estimation of baseline positions. Therefore, the proper selection of k is critical; yet there is no simple and reliable way for making a good choice. Another problem is that there is no obvious shoulder in density histograms for some words after discarding k highest density values. Figure 2.6(c) shows such an example. This problem is caused by ascenders and descenders of words. As a result, it is very difficult to adaptively choose suitable thresholds for baseline determination.

In this chapter, we will describe an algorithm based on the first derivative of the smoothed vertical density histogram $FD_{st}(0,j)(j \in H)$ of the slant and tilt-corrected image after horizontal line removal [41], which is equivalent to $FD_s(tilt,j)(j \in H)$ of the slant-corrected image after horizontal line removal, where H is the height of the image. As a consequence of horizontal line removal, this algorithm avoids the selection of k. The vertical density histogram is slant invariant, but it is very sensitive to ascenders and descenders of words. After tilt correction, we are able to remove this influence. This makes the shoulder in vertical density histogram more prominent (as shown Figure 2.6(e) compared with Figure 2.6(c)). The position of lower baseline is determined by the gradient of smoothed vertical histogram and the threshold:

$$y_l = \arg\max_{j \in H_l}\{FD_{st}(0,j) | D_{st}(0,j) > T_d\}, \tag{2.12}$$

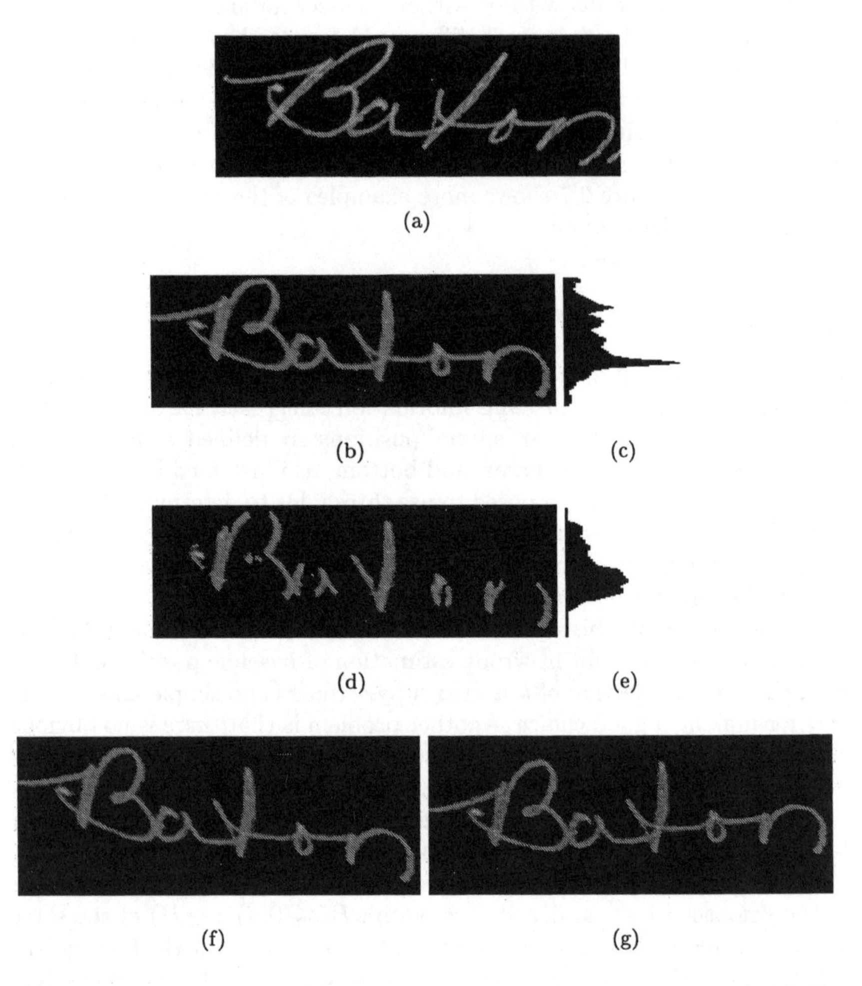

FIGURE 2.6. The influence of horizontal line removal on tilt correction. (a) The original image. In (b), the slant-corrected image is obtained from (a), where the estimated slant is 51°; in (c), the projected-density histogram is calculated from (b), where the estimated tilt is 3°; (d) the image of (b) after horizontal line removal; in (e), the projected-density histogram is calculated from (c), where the estimated tilt is −5°; (f) the tilt-corrected image of (b), where the tilt is estimated from (c); (g) the tilt-corrected image of (b), where the tilt is estimated from (e).

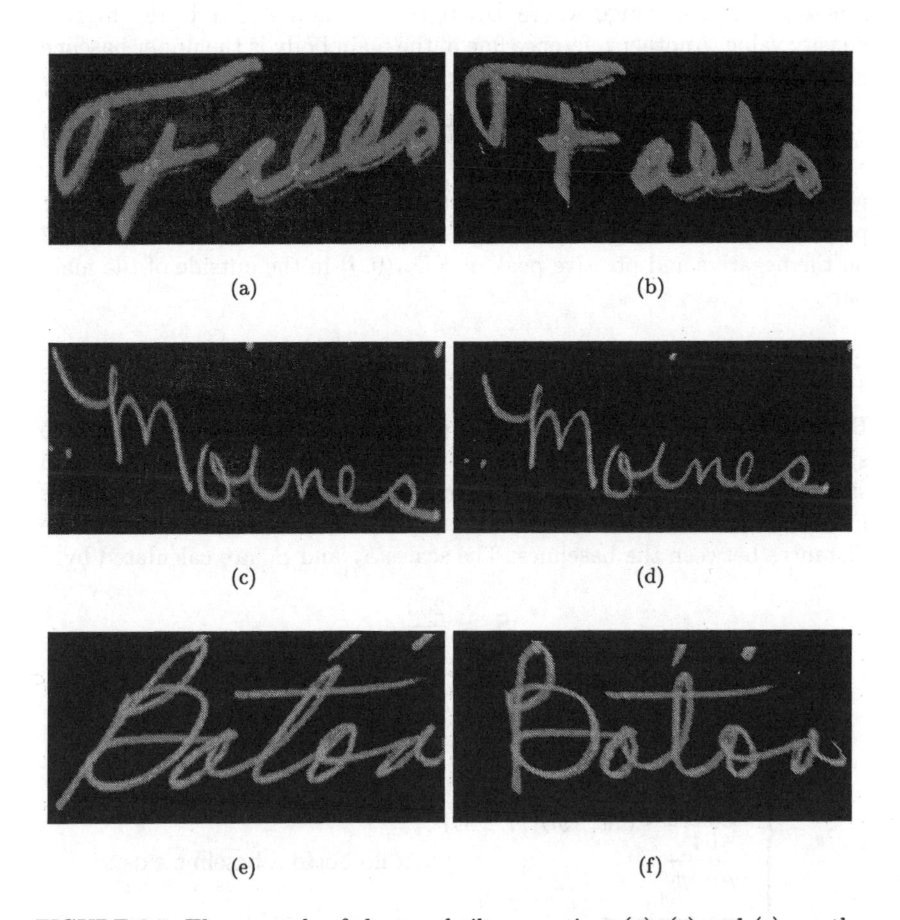

FIGURE 2.7. The example of slant and tilt correction. (a), (c) and (e) are the original images and (b), (d) and (f) are the slant and tilt-corrected images. In (a), the slant is 61° and tilt is 12.4°; in (c), the slant is 104° and tilt is −3.3°; in (e), the slant is 51° and tilt is 1.7°.

where

$$T_d = \min\{\bar{D}_{st}, MD_{st}/2\},$$

$D_{st}(0,j)$ is the smoothed vertical density histogram, T_d is the threshold, H_l is the range from the gravity center of an image down to the first point where $D_{st}(0,j) \leq T_d$, $\bar{D}_{st}$ is the average value of the vertical density function over the range where $D_{st}(0,j) > 0$, and MD_{st} is the highest density value. Another reference line of the main body is the upper baseline determined by locating negative peak in the histogram:

$$y_u = \arg\min_{j \in H_u} \{FD_{st}(0,j) | D_{st}(0,j) > T_d\}, \tag{2.13}$$

where H_u is the range from the gravity center of the image up to the first point where $D_{st}(0,j) \leq T_d$. The top and bottom baselines are found based on the negative and positive peak of $FD_{st}(0,j)$ in the outside of the main part.

Image Normalization

Because word sizes differ significantly, this makes it difficult to compare spatial positions of different words. We normalize images to a standard size of 200 × 128 pixels. In the normalized image, the lower baseline is located at $y_l = 64$. The scales are determined by the image width and the distances between the baselines. The scales S_x and S_y are calculated by

$$S_x = \frac{200}{IW},$$

$$S_y = \begin{cases} \dfrac{64}{\max\{y_t - y_l, y_l - y_b\}} & \text{if all baselines exist,} \\ \dfrac{64}{\max\{2 \times (y_u - y_l), y_l - y_b\}} & \text{if no top baseline exists,} \\ \dfrac{64}{y_t - y_l} & \text{if no bottom baseline exists,} \\ \dfrac{64}{2 \times (y_u - y_l)} & \text{if non of the above exists,} \end{cases}$$

where, IW is the image width, y_t, y_u, y_l and y_b are the y-axis coordinates of the top, upper, lower and bottom baselines (Figure 2.8).

Line Width Calculation

Line width of handwriting is useful to estimate optimal parameters of the Gabor filters, which are used for extracting features of handwriting. Usually, the line widths in a word are not uniform. To obtain all possible line widths in a word is impracticable, but we can estimate the average line

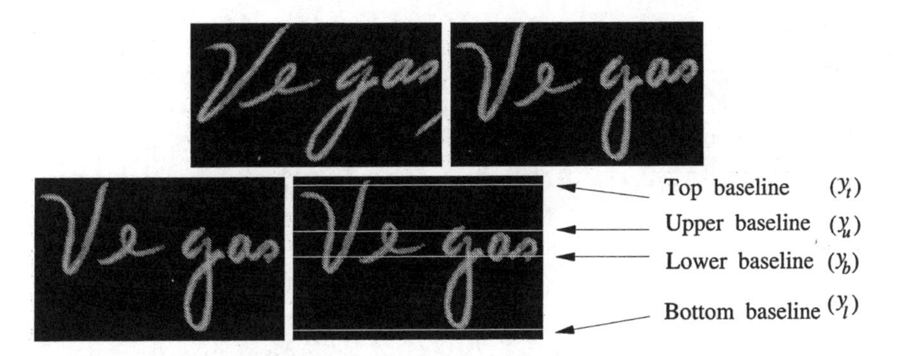

FIGURE 2.8. Top left: original image; Top right: slant corrected image Bottom left: slant and tilt corrected image; Bottom right: corrected image with four baselines.

width of a word. In the rest of this book, line width means the average line width. The algorithm for line width estimation is performed on the binarized word image. We assume that the length (L_l) of a line is much greater than the width (W) of the line, $L_l \gg W$. For an ideal line, the sum of its pixels is $S_{um} = L_l \times W$. After deleting its edge points, the sum of its pixels is $S'_{um} = L_l \times (W - 2)$, where $W \geq 2$. The width of the line can be easily calculated by

$$W = \frac{2 \times S_{um}}{S_{um} - S'_{um}}. \tag{2.14}$$

In real word images, a line can be viewed as an ideal line corrupted by noise. Figure 2.9 shows an example. In this example, an ideal line is corrupted by adding noise on only one side of its edges. For line width estimation, we propose two different methods to define edge points. One uses the neighbors of a pixel as its 8-connected points. The number of neighbors for pixel p is $N_8(p)$. Here, the edge point is defined as a nonzero point and its $N_8(p) \leq 6$, as shown in Figure 2.9(a). Because not every edge point is a *real* edge point, we call it pseudo-edge point. In the figure, '$\bullet$' represents a pseudo-edge point, and '$\star$' a nonzero pixel with its $N_8(p) \geq 7$. The estimated width of the line is 5. Another method uses the same definition of edge point as that used in slant correction shown in Figure 2.9(b) where the difference between Figures 2.9(a) and (b) is indicated by arrows. The estimated width of the line is still 5. For a real image illustrated in Figure 2.5(a), the average line widths estimated by the two methods are 7.82 and 7.86, respectively. This shows that Cai and Liu's method for line width estimation is quite robust to noise [38].

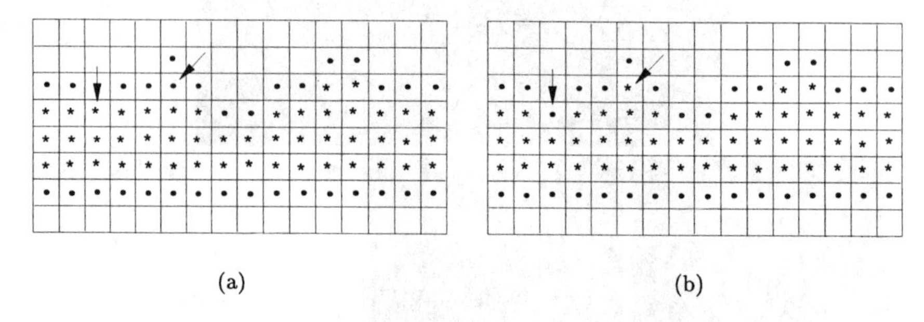

(a) (b)

FIGURE 2.9. Line width calculation. (a) uses the pseudo-edge definition; (b) uses four-connected neighbors to define edge point.

2.2 Feature Extraction from Binarized Images

In this section, we present some new, effective approaches to feature extraction from binarized images that have been used in most current handwriting recognition systems.

Skeletons

Skeletonization (thinning) is a process of data reduction and is used to represent patterns with elongated lines of unit width. There are numerous thinning algorithms available in the literature [150]. An ideal thinning algorithm should be an isotropic process that can be implemented in parallel. Unfortunately, most fully parallel algorithms cannot guarantee to preserve the connection of an image if 3×3 operations are used. Some line patterns of two pixels in width may vanish completely in such a process. Sequential thinning algorithms, which delete boundary points in the order of raster scans or contour following, are able to preserve connectivity. If the boundary points are examined for deletion in raster scans, a significantly asymmetric compression may be produced [220]. Better results can be obtained by thinning algorithms based on contour tracking and the elimination of insignificant extremities.

Outer Contours

Contour features are used in many handwritten-character recognition systems [139, 198, 197]. In Cai and Liu's HMM-based systems, features are extracted from closed outer contours that can be used in HMMs, because such features can be arranged into sequences that can accommodate the requirement of HMMs: the input signals must be sequential [37, 39]. A perfect contour must satisfy the following condition: If and only if every pixel of a contour has two neighboring pixels in the 8-neighbor system. This means that the line width must be at least 3-pixel wide.

Sequential features can be easily extracted around the perfect contours. Unfortunately, after pre-processing, the contours obtained directly from images cannot be reliably used to generate sequential features, because they are not perfect and have more spurious points, e.g., endpoints, cross-points, which make it difficulty to trace a contour; some points may have to be passed more than once. In order to obtain perfect contours, we use the following simple procedure:

1. Perform erosion once, which deletes all boundary pixels if they are deletable [150].
2. Perform standard dilation once [205].
3. Obtain the contour and delete redundant pixels.

Figure 2.10 shows examples of perfect contours.

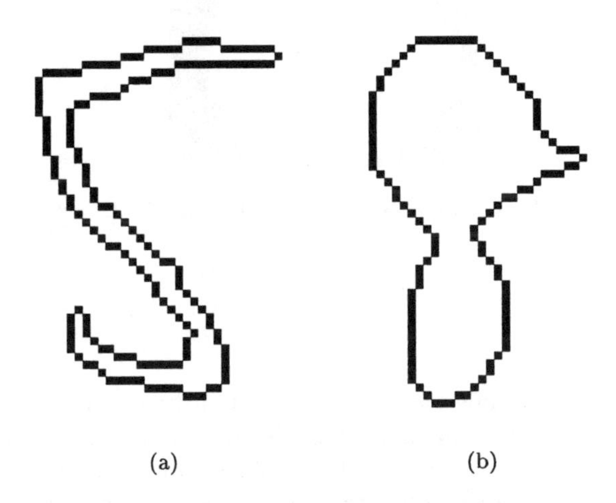

FIGURE 2.10. Examples of extracted contours.

2.3 Feature Extraction Using Gabor Filters

In the previous sections, we presented feature extraction from binarized images. However, extracting features from binary images does not fully use the available information such as the average line-segment width. As a consequence, algorithms based on such techniques can not differentiate the true line-segments from that induced by noise, nor can they differentiate holes filled by thick line-segments (shown in Figure 2.11) from a normal thick line-segment. Furthermore, binarization inevitably incurs information loss. More appropriate algorithms should be able to fully exploit the available information to produce features that better serve handwritten character recognition. In this section, we present a few novel algorithms for feature extraction directly from gray-scale images or pseudo gray-scale images based on orientated patterns obtained by using Gabor filters.

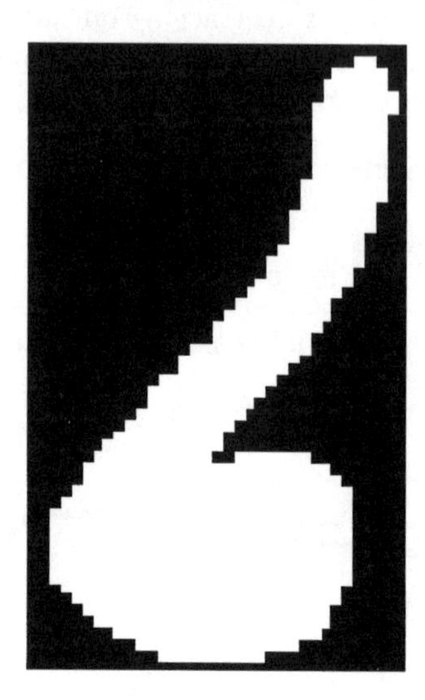

FIGURE 2.11. Example of a hole filled by thick line-segments.

Gabor Filter

In the 1940s, Gabor applied methods from quantum mechanics to the analysis of communications channels, and found that functions obtained by multiplying a complex exponential by a Gaussian function can achieve the smallest time (or space) bandwidth product. Such functions are a special kind of wavelets and known as the Gabor filters [90]. Gabor also showed that to distinguish the difference between two frequencies separated by Δf, the time interval between samples must be $\geq 1/\Delta f$, or $\Delta f \Delta t \geq 1$. That is, the product of signal's frequency bandwidth and time duration is limited by a lower bound. This is known as the Gabor Uncertainty Principle (which is a form of Heisenberg's Uncertainty Principle). This indicates that increasing localization of a signal in the time domain (decreasing Δt) will result in decreased localization in the frequency domain, and vice versa.

Marcelja was the first to use the Gabor filter to model the mammalian visual system [171]. He realized that odd or even symmetric receptive field profiles of simple cortical cells can be approximated by elementary Gabor functions in the spatial domain. He also observed that the visual scene was analyzed in terms of independent spatial frequency channels and that the cortical cells are tuned to specific spatial frequencies.

Daugman extended the original Gabor filter to a two-dimensional (2D) representation [67] (see (2.15)). Daugman showed that the 2D Gabor filter also satisfies the Gabor uncertainty principle. Jones evaluated Daugman's modeling of the visual cortex as a 2D Gabor filter, and concluded that the 2D Gabor filter can reasonably model most spatial aspects of the visual cortex [130]. Daugman also stated that a 2D Gabor representation is attractive for representing the response of the visual system as it simultaneously captures the fundamental properties of linear neural receptive fields: spatial localization, spatial frequency selectivity, and orientation selectivity [66].

Since then, Gabor filters have been applied to many areas in computer vision and image processing, including texture discrimination [75, 76, 83, 204], texture segmentation [23, 24, 57, 124], emergent frequency estimation [25], image compression [68], fractal dimension measurement [238], edge detection [175], motion estimation [108], shape-from-texture [237, 239], stereo disparity [219], object detection [44], text segmentation [125, 126], character recognition [93], and measurement [30, 164].

The Gabor filter is a typical wavelet and can be used to extract local information from image regions in both spatial and frequency domains, as opposed to global techniques such as the Fourier transform which extracts features that represent the properties of the entire image. The Gabor filter can extract such local features as orientations, surface roughness, and instantaneous frequencies of different textures. For these reasons it has been used extensively for texture analysis and segmentation. In such applications, the image is decomposed into a set of narrow-band spatial-frequency channels by a bank of Gabor filters whose responses are then analyzed. Multi-resolution segmentation is possible by filtering at different scales.

Heeger used a 3D Gabor filter in a space-time domain (x, y, t) to determine local image velocity in textured images [108]. The motion is considered as an orientation (plane) in the space-time domain. The Gabor filters are used to sample the power spectrum of the moving texture. The outputs of several filters are combined and the slant of the plane and the velocity of the moving texture are calculated. Sanger used the difference between the phase responses at corresponding points in two images to solve the correspondence problem in stereo disparity [219]. Garris *et al* used Gabor expansion coefficients as features for a neural network classifier to recognize numerals [93]. They stated that the use of Gabor filters reduced random noise and smoothed image structural irregularities. Mehrotra *et al* developed a step-edge detector based on odd Gabor filters [175]. The locations of the step edges were found from the local maximum in the absolute Gabor filter response. They showed that for optimum performance, a Gabor filter should be oriented perpendicular to the step edge with $\sigma = 1/\omega$, where σ is the width of the Gaussian and ω is the radial frequency of the sinusoid. They also observed that the local maxima are located at edge points and are independent of the scale of the Gaussian. Casasent *et al* used the Gabor filter to detect objects within scenes [44]: When the filter is centered on the object whose size equals the width of the Gaussian, the Gabor filter's response achieves its maximum value.

Liu *et al* developed an efficient and accurate analysis system for quantitative evaluation of collagen alignment in ligatures [164]. Further, Buse, Liu and Caelli developed a method for measuring physical parameters of lines by Gabor filters only [30].

In the above methods, a bank of Gabor filters with a range of frequencies and angles is applied to the image. The number of filters used varies from four to 96, and the parameters are selected on the basis of evenly distributing the filters in the frequency spectrum. In our applications, we use the response from each of the filters separately as each filter is tuned to extract only parts of a word oriented at a specific angle.

The Gabor filter used in the context of handwriting recognition is defined in the spatial domain $g(x,y)$ and the frequency domain $G(u,v)$. In both domains, the Gabor filter is a sinusoidal plane wave of a given orientation and frequency within a 2-D Gaussian envelop, which can be formally stated as follows:

$$g(x,y) = \exp\{-\pi(\frac{x'^2}{\sigma_x^2} + \frac{y'^2}{\sigma_y^2})\}\exp\{\jmath 2\pi(u_0 x + v_0 y)\}, \quad (2.15)$$

and its 2-D Fourier transforms $G(u,v)$ is:

$$G(u,v) = \exp\{-\pi[(u' - u_0')^2\sigma_x^2 + (v' - v_0')^2\sigma_y^2]\}, \quad (2.16)$$

where,

$$\begin{aligned} x' &= x\cos\phi + y\sin\phi & y' &= -x\sin\phi + y\cos\phi \\ u' &= u\cos\phi + v\sin\phi & v' &= -u\sin\phi + v\cos\phi \\ u_0' &= u_0\cos\phi + v_0\sin\phi & v_0' &= -u_0\sin\phi + v_0\cos\phi \\ u_0 &= f\cos\theta & v_0 &= f\sin\theta \end{aligned}$$

and $\jmath = \sqrt{-1}$. Eq.(2.16) is the equation of a 2D Gaussian function centered at (u_0, v_0) in the frequency domain. It is a 2D band-pass filter. The parameters σ_x and σ_y determine the frequency and orientation bandwidths of the filter. The position (f, θ; or u_0, v_0), direction (ϕ, with respect to the u (or x) axis), and aspect ratio (σ_x, σ_y) are the variables.

The effect of these five parameters on the shape of the Gabor filter is shown for the spatial and frequency domains in Figures 2.12 and 2.13. Figure 2.12(a) shows the response with no rotations ($\phi = \theta = 0°$), and thus will be used as a reference to which the other subfigures are referred in order to show the effect of each of the parameters. It also has an aspect ratio (σ_x/σ_y) of 2 which was chosen to enhance the visual effects in the figures.

Figure 2.12(b) shows the effect of rotating the orientation θ by 45° in a counterclockwise direction. The parameter θ defines the angle of the wave form, and is referred to as the *orientation* of the Gabor wavelet. This is in contrast to the effect of the parameter ϕ shown in Figure 2.12(c). Here, the axis on which the filter lies has been rotated by 45° with respect to the X-axis. The angle of the waveform is unchanged, but the relative direction between the peaks of the oscillations has been rotated, and ϕ is thus referred to as the *direction* of the filter. It can also be interpreted as the direction of the major axis of the 2D Gaussian envelope. The size of the Gaussian envelope is determined by the parameters σ_x and σ_y. Finally, Figure 2.12(d) shows the effect of lowering the frequency f only. As the size of the Gaussian envelope is unchanged, the number of oscillations contained within the same area is lower as compared to Figure 2.12(a).

The effect of the parameter variations within the frequency domain is shown in Figure 2.13. Note that the shape of the Gabor filter, as described in Eq.(2.16) is that of a 2D Gaussian whose position, size, and orientation are determined by the values of its parameters. Again, all comparative references are made with respect to Figure 2.13(a). The position of the center of the Gaussian envelope within the frequency plane is determined by θ and f, forming a polar coordinate pair within the U-V plane, i.e., $u = f \cos\theta, v = f \sin\theta$. This can be seen in Figures 2.13(b) and 2.13(d), respectively. Again, the direction of the Gaussian is determined by ϕ, (Figure 2.13(c)). The Gaussian has been rotated by 45° counterclockwise from Figure 2.13(a). In the frequency domain, an increase in σ_x and σ_y results in a decrease in the spread of the Gaussian.

Daugman formally proved that the joint 2-D resolution for this family of filters achieves the theoretical limit of $1/16\pi^2$ regardless of the values of the parameters [67]. This means that an increase of the filter spread in the X-axis will result in a decrease of the frequency bandwidth in u-axis and vice versa. There is the same relation in y and v. Figure 2.14 shows the uncertainty relation in the space and frequency domains. Daugman also showed some examples of experimentally measured 2-D receptive-field profiles of simple cells in cat visual cortex as compared with the best fitting 2-D Gabor functions and showed that the 2-D Gabor filter can reasonably model the 2-D receptive field profiles of simple cells in mammalian visual cortex.

The Gabor filter has the following major properties:

- It is tunable to specific orientations. This allows us to extract the features of line-segments at any possible orientation.
- Its orientation bandwidth is adjustable. So, we can use the least number of Gabor filters to achieve the required orientational accuracy.
- It optimizes the general uncertainty in both space and frequency domains.
- It can extract local information from images. This property is useful to obtain the local orientation of a curve.

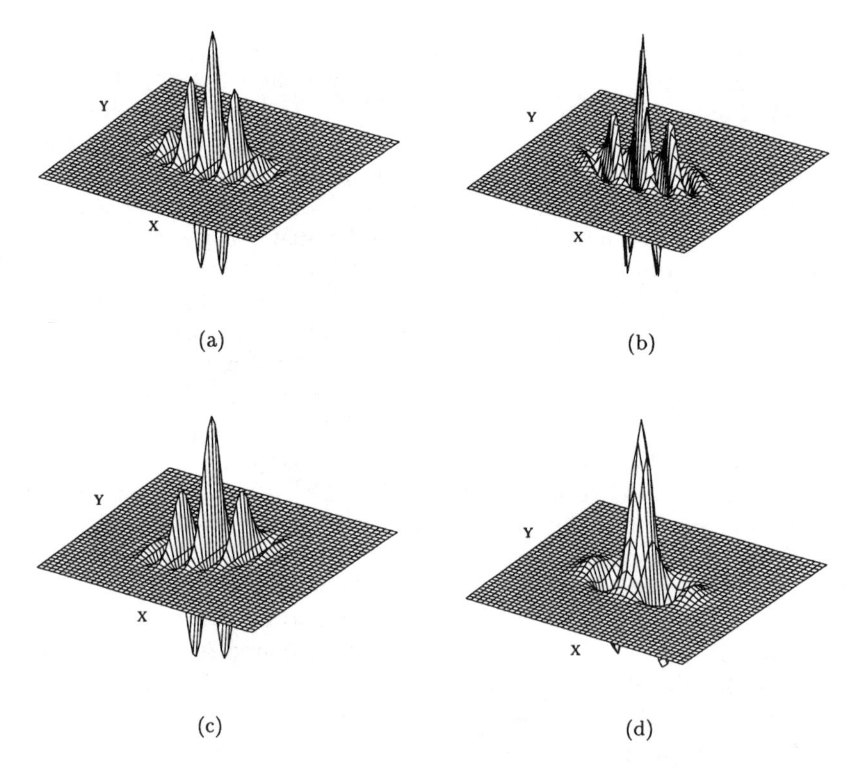

FIGURE 2.12. Effects of the Gabor filter parameters, displayed in the spatial domain. (a) $\theta = 0°$, $\phi = 0°$, $\sigma_x = 0.1$, $\sigma_y = 0.05$, $f = 25$; (b) $\theta = 45°$; (c) $\phi = 45°$; (d) $f = 10$.

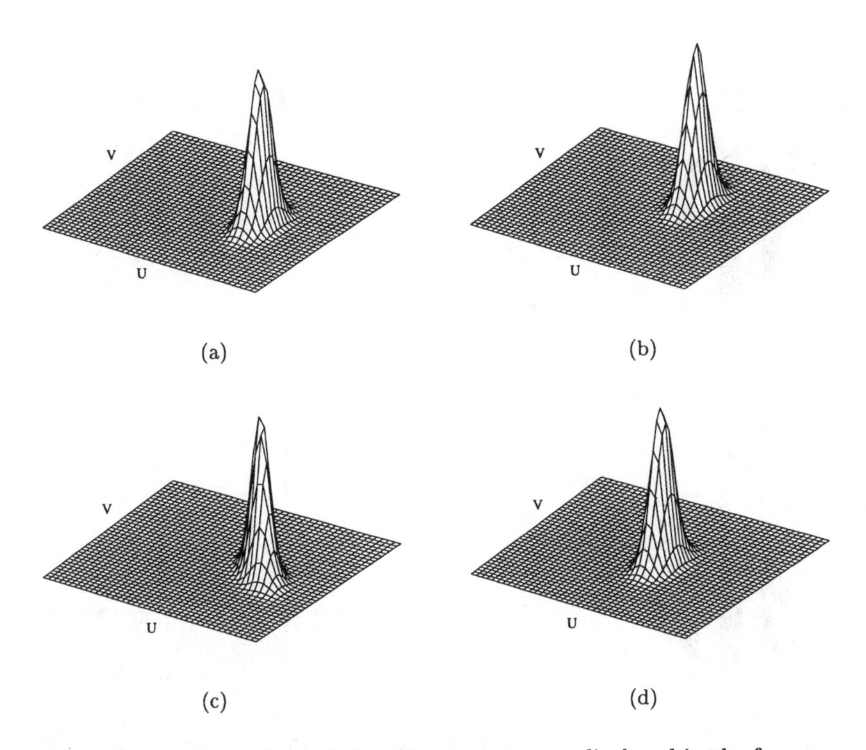

FIGURE 2.13. Effects of the Gabor filter parameters, displayed in the frequency domain. (a) $\theta = 0°$, $\phi = 0°$, $\sigma_x = 0.1$, $\sigma_y = 0.05$, $f = 25$; (b) $\theta = 45°$; (c) $\phi = 45°$; (d) $f = 10$. (Here, only the positive frequencies are shown as the negative frequencies are symmetric about the origin and simply duplicate the information.)

- The output of the filter is robust to noise, because the Gabor filter uses the information of all pixels within the kernel.
- It can be implemented by optoelectronic processor at a high speed. Because the Gabor filter is not steerable, usually, the operator needs a large number of additions and multiplications. However, Gabor filtering operations are particularly easy to be implemented by an optical system with VLSI-based processors [246].

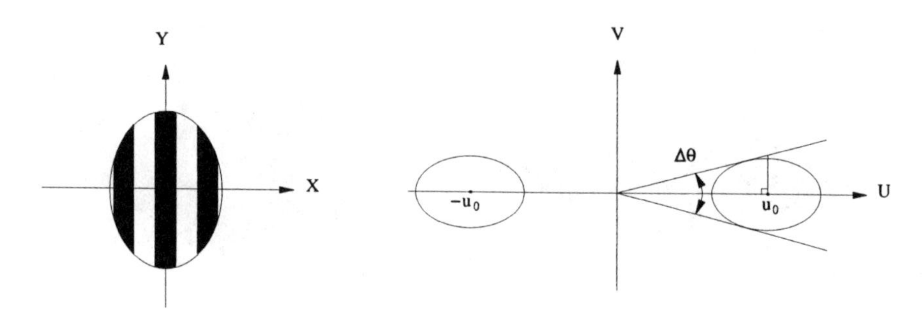

(a) Spatial wave pattern at 90°

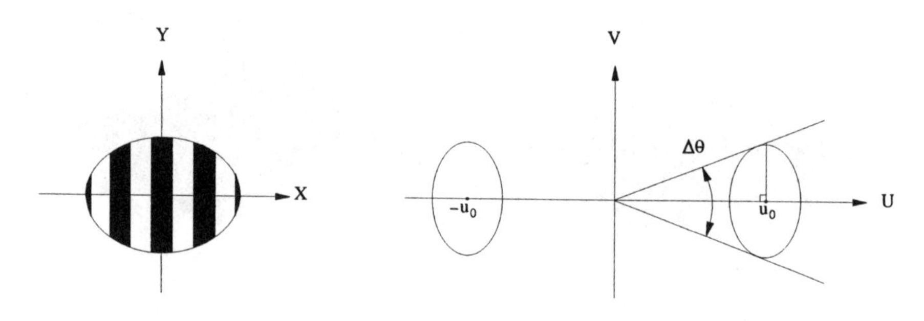

(b) Spatial Wave pattern at 0°

FIGURE 2.14. The uncertainty relation in space and frequency domains.

In this book, we use the Gabor filter to extract the features of handwritten characters and words.

A: Determination of Gabor Filter Parameters

Correctly determining the Gabor filter parameters is essential for consistent and meaningful feature extraction in most applications [254, 186]. In [186], some filter parameters are preset. Some authors selected the parameters by trial and error. These methods cannot effectively deal with variations in handwriting. Recently, Cai and Liu developed a new method to estimate the parameters of the Gabor filter for extracting oriented line-segments [38]. In this section, we modify the method presented in [38] to compute the parameters of the Gabor filter, which adapt to different lengths and widths of line-segments in handwritten characters.

B: Orientation Bandwidth

We extract features with a group of Gabor filters. The number of Gabor filters is determined by the orientation bandwidth at half of the maximum power response in the 2D frequency domain. According to orientation bandwidths of cortical cells, the mean bandwidths cover the range from 26° to 39° [67]. Therefore, the least number of Gabor filters covering the range [0°, 180°] is from 5 to 7. For convenience, we choose $n = 8$. Our experimental results show that this is adequate. The orientation bandwidth $\Delta\theta$ of Gabor filters can be set to 22.5° accordingly. In order to reconstruct the original word from extracted parts without broken lines, we set $\Delta\theta = \alpha \times 22.5°$ for $n = 8$, where $1.5 \geq \alpha > 1$.

C: Frequency

The frequency f determines the 2-D spectral centroid positions of the Gabor filter. This parameter is derived with respect to the average width of lines in handwritten words, where the average width of lines can be estimated by the method proposed in [38]. Let us consider an ideal rectangular pulse line $\mathcal{L}(x, y)$ whose length is much greater than its width, where $\mathcal{L}(x, y) = u(x + W/2) - u(x - W/2)$, $u(\cdot)$ is the step function, and W is the line width. Figure 2.15(a) shows such a line. If f is set too high, the Gabor filter will produce two peaks at edges of the rectangular pulse line. In order to produce a single peak for the given line, the output of the Gabor filter must satisfy the following conditions:

$$Out_{gr}(0, s) \geq Out_{gr}(t, s) \qquad \frac{W}{2} \geq t > 0, \tag{2.17}$$

$$Out_{gr}(t_0, s) \geq Out_{gr}(t_1, s) \qquad \frac{W}{2} \geq t_1 > t_0 > 0, \tag{2.18}$$

where the output of the Gabor filter can be represented as convolution between $g_r(x, y)$ and $\mathcal{L}(x, y)$:

$$Out_{gr}(t, s) = \int_{-\infty}^{\infty} \int_{-\infty}^{\infty} g_r(x, y) \mathcal{L}(x - t, y - s) \mathrm{d}x \mathrm{d}y, \tag{2.19}$$

where $g_r(x, y)$ can be derived from (2.15) by setting the orientation to the angle of the line as follows:

$$g_r(x, y) = \exp\{-\pi(\frac{y^2}{\sigma_x^2} + \frac{x^2}{\sigma_y^2})\} \cos(2\pi f x). \tag{2.20}$$

By substituting $g_r(x, y)$ into (2.19), we obtain

$$Out_{gr}(t,s) = \int_{t-\frac{W}{2}}^{t+\frac{W}{2}} \int_{-\infty}^{\infty} \exp\{-\pi(\frac{y^2}{\sigma_x^2} + \frac{x^2}{\sigma_y^2})\} \cos(2\pi f x) \mathrm{d}x\mathrm{d}y. \quad (2.21)$$

Because $Out_{gr}(t, s)$ is differentiable, (2.17) and (2.18) can be expressed by

$$\begin{array}{ll} \frac{\mathrm{d}Out_{gr}(t,s)}{\mathrm{d}t} \leq 0 & \frac{W}{2} \geq t \geq 0, \\ \frac{\mathrm{d}Out_{gr}(t,s)}{\mathrm{d}t} \geq 0 & 0 > t \geq -\frac{W}{2}. \end{array} \quad (2.22)$$

Thus, we obtain

$$\exp\{-\pi\frac{(W/2+t)^2}{\sigma_y^2}\} \cos[2\pi f(W/2+t)] - \exp\{-\pi\frac{(-W/2+t)^2}{\sigma_y^2}\} \cos[2\pi f(-W/2+t)] \leq 0, \quad \frac{W}{2} \geq t \geq 0. \quad (2.23)$$

Because σ_y can be any positive value, (2.23) must hold regardless of the parameter σ_y. Therefore, if

$$\cos[2\pi f(-W/2+t)] \geq \cos[2\pi f(W/2+t)] \qquad \frac{W}{2} \geq t \geq 0 \quad (2.24)$$

holds, so does (2.23). (2.24) is equivalent to

$$\sin \pi f W \cdot \sin 2\pi f t \geq 0 \qquad \frac{W}{2} \geq t \geq 0. \quad (2.25)$$

Since f is positive, from (2.25) we obtain

$$0 < f \leq 1/W. \quad (2.26)$$

Let us consider another case. If two lines are very close or touch each other as shown in Figure 2.15(b), they can be viewed as one line with $2W$ width. A good feature extractor should be able to differentiate it from a line with width W. Therefore, the output of the Gabor filter should display two peaks instead of one. According to (2.26), in order to produce two peaks in this case, the parameter f must be

$$f > \frac{1}{2W}. \quad (2.27)$$

By combining (2.26) and (2.27), we get a suitable f given by

$$0.5/W < f \leq 1/W, \quad (2.28)$$

which can produce the desired output. Therefore, we set $f = \beta/W$, where β is in the range of [0.5, 1].

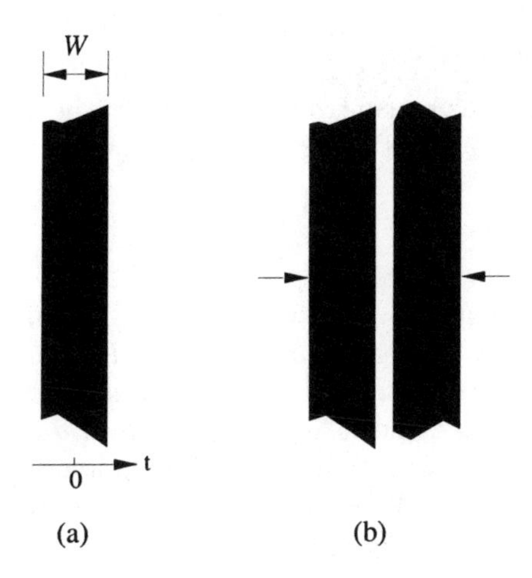

FIGURE 2.15. Examples of an ideal line and two lines close to each other.

D: The Standard Deviation σ_x

The standard deviation σ_x determines the spread of Gabor filter in the ϕ direction. The orientation bandwidth is also mainly determined by σ_x and f. The relationship between orientation bandwidth (in radians) and f and σ_x is illustrated in Figure 2.14. If $\Delta\theta/2 \ll \pi$ ($\Delta\theta = \Delta\phi$), the relationship among them can be approximated by

$$\Delta\theta = \alpha\frac{\pi}{n} \approx 2\arctan(\frac{\Delta F_\phi/2}{f}), \tag{2.29}$$

where ΔF_ϕ is defined as the 3-dB frequency bandwidth of the filter in v direction when $\phi = \pi/2$. From (2.16), we obtain

$$G(u,v)|_{\phi=\pi/2} = exp\{-\pi[v^2\sigma_x^2 + (u-u_0)^2\sigma_y^2]\}. \tag{2.30}$$

By applying the condition of 3-dB frequency bandwidth to (2.30), we have

$$G(u_0, \frac{\Delta F_\phi}{2})|_{\phi=\pi/2} = exp\{-\pi(\frac{\Delta F_\phi}{2}\sigma_x)^2\} = \frac{\sqrt{2}}{2}. \tag{2.31}$$

This gives

$$\Delta F_\phi = \frac{\lambda}{\sigma_x}, \tag{2.32}$$

where $\lambda = \sqrt{2\ln 2/\pi}$. Because the ratio between π and half orientation bandwidth is $\frac{\pi}{\Delta\theta/2} = \frac{2n}{\alpha} \geq n$ and $n = 8 \gg 1$, (2.29) holds. Therefore, from the above equations, we obtain

$$\sigma_x \approx \frac{n\lambda W}{\alpha\beta\pi}. \tag{2.33}$$

E: The Standard Deviation σ_y

The standard deviations σ_y controls the spread of the Gabor filter in the θ direction. Similar to (2.32), the relationship between σ_y and ΔF_θ is given by

$$\Delta F_\theta = \frac{\lambda}{\sigma_y}, \tag{2.34}$$

where ΔF_θ is the frequency bandwidth of the filter in u direction when $\theta = 0$. This means that decreasing σ_y will result in a larger spectral spread in frequency domain. Because as the width of a line-segment decreases, its spectral spread increases, the relationship between σ_y and ΔF_θ is similar to that between the line width and its spectral spread and σ_y should be a function of the line width W. As σ_x is indirect proportion to W in (2.33), we may express σ_y in terms of σ_x:

$$\sigma_y = k\mathcal{F}(\sigma_x), \tag{2.35}$$

where k is a constant. This relationship is consistent with the constraint on the kernel shape of cortical cells [67] and can be approximated as $\mathcal{F}(\sigma_x) = \sigma_x$. Thus we have

$$\sigma_y = k\sigma_x. \tag{2.36}$$

For cortical cells, the ratio σ_x/σ_y is independent of f and is constrained to a small range, whereas the product $\sigma_x\sigma_y$ varies over a much bigger range.

However, the selection of k is important to feature extraction. If k is small, the difference between the outputs of the Gabor filter at different orientations will be small. This may result in errors in orientation estimation. Furthermore, the outputs of the filter with small k are sensitive to noise because there are too few pixels within the kernel, on which the outputs depend. On the other hand, a large k will result in a strong interference between the outputs of the filter for two lines that are close to each other. Unfortunately, we have little clue as to how to determine the coefficient k from (2.15) and (2.16). A reasonable way of selecting k may be based on the positive correlation among kernel shapes of cortical cells. It was reported in [67] that the space-domain measurements of k in populations of simple cells usually range between 0.25 and 1. After examining the feature extraction results over this range, we find it is appropriate to set $k = 1$ for binary images and $k = 0.85$ for gray-scale images.

2.3.1 Skeletonization Using Gabor Filters

In this section, we present a new thinning method using the Gabor filter. This thinning method includes the following steps:

- calculating the responses of the real and imaginary parts of the Gabor filter in eight directions,

- determining the principal directions for each ridge pixel,
- extracting skeletons of images according to the principal directions, and
- refining the skeletons of handwritten images.

For binary images, there is an extra step: converting binary images into pseudo gray-scale images.

A: Calculation of the Responses of the Gabor Filter

The responses of the real and imaginary parts of the Gabor filter in the eight directions can be calculated by

$$Out_{gr}(x_i, y_i)_\eta = \int_{-\infty}^{\infty}\int_{-\infty}^{\infty} g_r(x,y)_\eta Z(x - x_i, y - y_i)\mathrm{d}x\mathrm{d}y, \tag{2.37}$$

$$Out_{gi}(x_i, y_i)_\eta = \int_{-\infty}^{\infty}\int_{-\infty}^{\infty} g_i(x,y)_\eta Z(x - x_i, y - y_i)\mathrm{d}x\mathrm{d}y, \tag{2.38}$$

where $Z(x,y)$ is the gray value of a pixel at (x,y), $g_r(x,y)_\eta$ and $g_i(x,y)_\eta$ are the real and imaginary parts of Gabor filter in the orientation, $\phi = \phi_\eta$, where ϕ_η shown in Figure 2.16 is defined as

$$\phi_\eta = \eta\pi/8 \qquad 0 \leq \eta \leq 7. \tag{2.39}$$

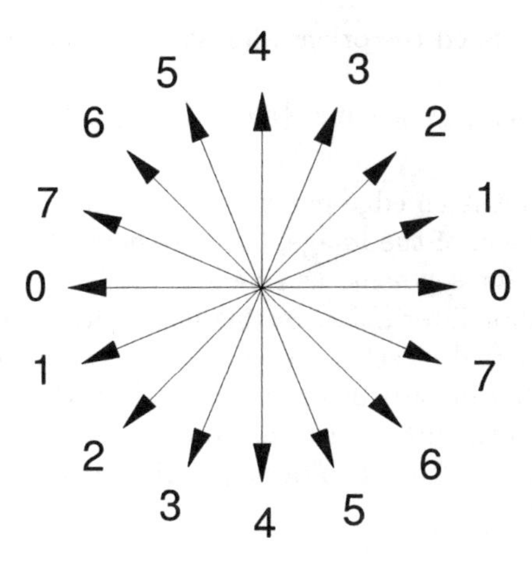

FIGURE 2.16. The eight orientations of Gabor filter ϕ_η.

For a digital image, (2.37) and (2.38) become

$$Out_{gr}(x_i, y_i)_\eta = \sum_x \sum_y g_r(x,y)_\eta Z(x - x_i, y - y_i)\mathrm{d}x\mathrm{d}y, \tag{2.40}$$

$$Out_{gi}(x_i, y_i)_\eta = \sum_x \sum_y g_i(x, y)_\eta Z(x - x_i, y - y_i) \mathrm{d}x\mathrm{d}y. \tag{2.41}$$

In practice, the range of summation is quite small. Since $\exp(-\pi 1.5^2) < 1/1000$ and $\sigma_x \geq \sigma_y$, the range of summation can be limited to $r < 1.5\sigma_x$ without losing much accuracy, where $r = \sqrt{(x - x_i)^2 + (y - y_i)^2}$. The responses of the Gabor filter are robust to noise, because they depend on all pixels over the kernel. Figure 2.17 illustrates the robustness of $Out_{gr}(x_i, y_i)_\eta$ to noise, where Figure 2.17(D) and Figure 2.17(E) are the responses $Out_{gr}(x_i, y_i)_4$ for the inputs shown in Figure 2.17(B) and Figure 2.17(C). In addition, a larger value of $Out_{gr}(x_i, y_i)_\eta$ means that there is a higher probability that the pixel at (x_i, y_i) is a ridge point of a line with orientation $\eta\pi/8$, and a larger value of $|Out_{gi}(x_i, y_i)_\eta|$ means that there is a higher probability that the pixel is an edge point of the line. Therefore, the responses of the Gabor filter can be used to determine the principal directions at a given pixel's location.

B: Determination of Principal Directions

A typical line consists of ridge-like points and edge points. In this book, a ridge-like point is defined as a point at (x_i, y_i) which satisfies the following condition:

$$f_m(x_i, y_i) = \max_\eta \{f(x_i, y_i)_\eta\} > 0 \qquad 0 \leq \eta \leq 7, \tag{2.42}$$

where $f(x_i, y_i)_\eta$ is called the orientation strength function defined by

$$f(x_i, y_i)_\eta = Out_{gr}(x_i, y_i)_\eta - \max_\mu \{|Out_{gi}(x_i, y_i)_\mu|\} \qquad 0 \leq \mu \leq 7. \tag{2.43}$$

Otherwise, the point is an edge point. Because edge points have no contribution to the skeleton of the image, they are deleted.

Orientations of line-segments can be accurately measured using the responses of the Gabor filter [254]. In this book, the Gabor filter is used to determine the principal directions of a ridge-like point. For each ridge-like point, there are one or two principal directions. Usually, each ridge-like point has only one principal direction. However, at a T-cross or an X-cross, the ridge-like point will have two principal directions, which are defined as follows:

$$d_1(x_i, y_i) = \arg\max_\eta \{f(x_i, y_i)_\eta\} \qquad 0 \leq \eta \leq 7, \tag{2.44}$$

$$d_2(x_i, y_i) = \arg\max_\mu \{f(x_i, y_i)_\mu\}, \tag{2.45}$$

where $d_1(x_i, y_i)$ is the first principal direction with the maximum orientation strength function, $d_2(x_i, y_i)$ is the second principal direction with

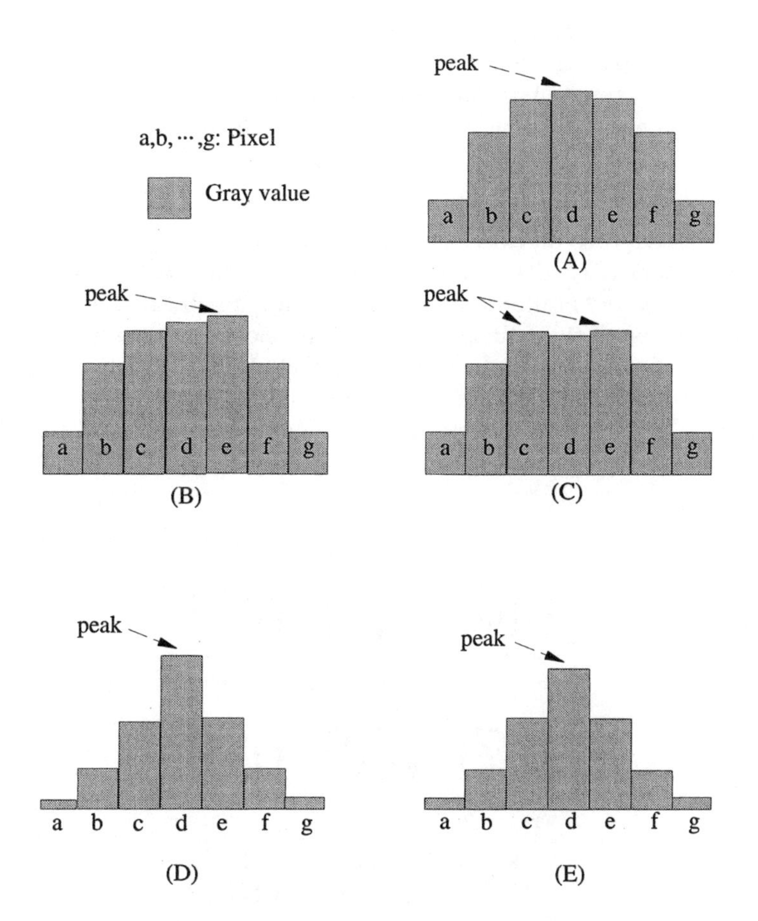

FIGURE 2.17. Examples of the robustness of the responses to noise.

the largest local maximum, $\mu \neq d_1(x_i, y_i)$, and $f(x_i, y_i)_\mu$ must be a local maximum. Therefore, $f(x_i, y_i)_\mu$ must satisfy the following two conditions:

$$f(x_i, y_i)_\mu > f(x_i, y_i)_{\mu+1},$$

$$f(x_i, y_i)_\mu > f(x_i, y_i)_{\mu-1},$$

where $f(x_i, y_i)_8 = f(x_i, y_i)_0$ and $f(x_i, y_i)_{-1} = f(x_i, y_i)_7$. If we extracted only one principal direction, the extracted skeleton of the image would be broken at the cross points. Therefore, it is necessary to use two principal directions to perform skeletonization instead of using one.

C: Skeletonization

In previous research on extracting feature directly from gray-scale images, skeletonization is carried out by heuristic rules that are established according to the orientation information. Wang and Pavlidis' method uses the information of the estimated first and second directional derivatives to determine whether or not to delete a pixel [251]. Lee and Kim's method makes the decision according to the rules based on directional zero-crossing conditions [156]. In this section, we perform skeletonization according to a set of rules based on the orientation strength functions in the 3×3 window as shown in Figure 2.18.

p[8]	p[1]	p[2]
p[7]	p[0]	p[3]
p[6]	p[5]	p[4]

FIGURE 2.18. Pixel $p[0]$ and its neighbors.

Let $p[0]$ denote the binary value at (x_i, y_i) after skeletonization, where $p[0]$ takes the value of 0 (the pixel is deleted) or 1. Let $f_m[0]$ be the maximum of orientation strength functions at (x_i, y_i), $f[0]_{d_2}$ be the orientation strength function at (x_i, y_i) in the direction d_2, $f[1]_{d_2}$ be the orientation function at (x_i, y_i+1) in the direction d_2 and so on. The heuristic decision-making rules are as follows:

Rule 1: $d_1 = 0, 1$ or 7, or $d_2 = 0, 1$ or 7.

1.1) If $f_m[0] + f_m[3] + f_m[7] \geq f_m[1] + f_m[2] + f_m[8]$ & $f_m[0] + f_m[3] + f_m[7] \geq f_m[4] + f_m[5] + f_m[6]$, $p[0] = 1$.

1.2) If $f[0]_{d_2} + f[3]_{d_2} \geq f_m[1] + f_m[2]$ & $f[0]_{d_2} + f[3]_{d_2} \geq f_m[4] + f_m[5]$ or $f[0]_{d_2} + f[7]_{d_2} \geq f_m[1] + f_m[8]$ & $f[0]_{d_2} + f[7]_{d_2} \geq f_m[5] + f_m[6]$, $p[0] = 1$.

Rule 2: $d_1 = 3, 4$ or 5, or $d_2 = 3, 4$ or 5.

2.1) If $f_m[0] + f_m[1] + f_m[5] \geq f_m[2] + f_m[3] + f_m[4]$ & $f_m[0] + f_m[1] + f_m[5] \geq f_m[6] + f_m[7] + f_m[8]$, $p[0] = 1$.

2.2) If $f[0]_{d_2} + f[1]_{d_2} \geq f_m[2] + f_m[3]$ & $f[0]_{d_2} + f[1]_{d_2} \geq f_m[7] + f_m[8]$ or $f[0]_{d_2} + f[5]_{d_2} \geq f_m[3] + f_m[4]$ & $f[0]_{d_2} + f[5]_{d_2} \geq f_m[6] + f_m[7]$, $p[0] = 1$.

Rule 3: $d_1 = 1, 2$ or 3, or $d_2 = 1, 2$ or 3.

3.1) If $f_m[0] + f_m[2] + f_m[6] \geq f_m[3] + f_m[4] + f_m[5]$ & $f_m[0] + f_m[2] + f_m[6] \geq f_m[1] + f_m[7] + f_m[8]$, $p[0] = 1$.

3.2) If $f[0]_{d_2} + f[2]_{d_2} \geq f_m[3] + f_m[4]$ & $f[0]_{d_2} + f[2]_{d_2} \geq f_m[1] + f_m[8])$ or $f[0]_{d_2} + f[6]_{d_2} \geq f_m[4] + f_m[5]$ & $f[0]_{d_2} + f[6]_{d_2} \geq f_m[7] + f_m[8]$, $p[0] = 1$.

Rule 4: $d_1 = 5, 6$ or 7, or $d_2 = 5, 6$ or 7.

4.1) If $f_m[0] + f_m[4] + f_m[8] \geq f_m[1] + f_m[2] + f_m[3]$ & $f_m[0] + f_m[4] + f_m[8] \geq f_m[5] + f_m[6] + f_m[7]$, $p[0] = 1$.

4.2) If $f[0]_{d_2} + f[8]_{d_2} \geq f_m[1] + f_m[2]$ & $f[0]_{d_2} + f[8]_{d_2} \geq f_m[6] + f_m[7]$ or $f[0]_{d_2} + f[4]_{d_2} \geq f_m[2] + f_m[3]$ & $f[0]_{d_2} + f[4]_{d_2} \geq f_m[5] + f_m[6]$, $p[0] = 1$.

Rule 5: If $p[0]$ is not set to 1 by above rules, $p[0] = 0$.

The first four rules are based on the same principles for the four directions. The measures to preserve the connectedness of skeletons have been taken into consideration in these rules. Figure 2.19 illustrates the process of preservation of connectedness of this method, where the gray-scale image is obtained by convoluting a Gaussian function with $\sigma = 1.0$ with two orthogonal lines having uniform intensity (e.g., $Z(x, y) = 100$) and five pixels in width shown in Figure 2.19(a). The orientation strength functions are extracted by the Gabor filter, where the parameters of the Gabor filter are: $\sigma_x = 6$, $\sigma_y = 6$, and $f = 0.12$. For peak and ridge points, if the principal directions are near parallel with x-axis, the orientation strength functions satisfy the conditions in *Rule 1.1*; and these points will not be deleted. For saddle points, they should be used as connectors for poorly printed text images and used as separators for high quality text images [251]. However, the quality of images is difficult to be determined automatically. In this method, if $f[0]_{d_1} \leq 0$ and $f[0]_{d_2} \leq 0$, this saddle point will be deleted; otherwise the saddle point will be kept because its strength function satisfies *Rule 1.1*, e.g., the cross-points in Figures 2.19(b) and (c). For pit-like points, they are usually deleted, but if they are very shallow and their orientation strength functions satisfy *Rule 1.2* as the right neighbor of the cross point in Figure 2.19(c), they will not be deleted. Therefore, we are able to preserve the connectivity. If the decision were made based on $f_m[0] \geq f_m[1]$ and $f_m[0] \geq f_m[5]$, the pit-like point in Figure 2.19(c) would be deleted. These measures are useful to alleviate the problem of broken lines, especially at crosses and points with large curvatures.

In real-world applications, both the line width and intensity may vary in the image field. These variations at joints are the main causes of broken lines. If the intensity of the horizontal line is 60 and the width of lines remains unchanged as shown in Figure 2.19(a), our method can still preserve the connectivity as shown in Figure 2.19(c). When the intensity of the horizontal line changes to 59 or lower, our method will produce two separated skeleton lines. If the intensity remains unchanged and the width of the horizontal line is reduced to one or two pixels, the connectivity cannot be preserved. In the next example, we change both the intensity and width of the horizontal line. The width of the horizontal line is fixed to three pixels and the intensity of the line keeps decreasing from 100. We find that our method can preserve connectivity until the intensity reaches 68, which represents 40% and 32% variations in line width and intensity, respectively.

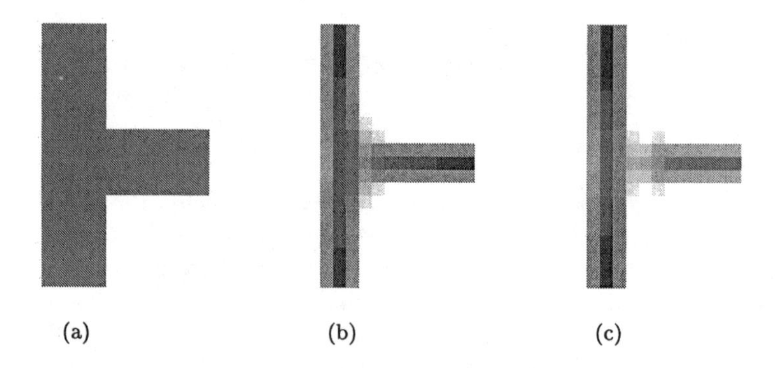

FIGURE 2.19. Examples of preservation of connectedness.

Perfect skeletons have unit width. However, the results obtained by the initial skeletonization may contain redundant pixels. Therefore, further refining of the skeletons is needed, for which we use the conventional thinning algorithms that introduce no significant distortions [150, 21].

D: Image Conversion

When binarization is performed on gray-scale handwritten images, edges in these images are sensitive to noise. In addition, binarization may produce broken or touching line-segments. Therefore, some distortions exist in binary images. These distortions can be greatly alleviated if the average line width is taken into consideration. In general, most skeleton points are located at the middle of the line; therefore, the pixels with distances about $W/2$ should play an important role in skeletonization. As the distortions are more likely to occur at edges, the edge points play a lesser role in skeletonization. In our method, we convert the binary images into pseudo gray-scale images, which uses the pseudo gray-scale image conversion as illustrated in Figure 2.20. The pixels with distances about $W/2$ are set to higher intensity, whereas the pixels with distances far from $W/2$ are set to lower intensity, where the distance can be computed by the distance transform technique [255]. The next steps of thinning for binary images are the same as those used for gray-scale images.

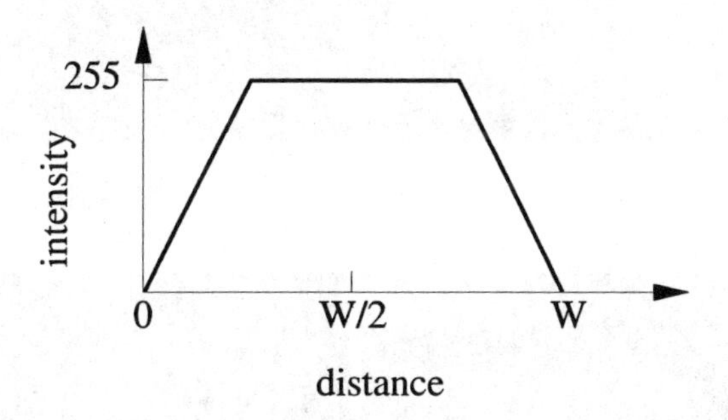

FIGURE 2.20. The relationship between intensity and distance.

2.3.2 Results of Skeletonization

A: Database

To evaluate our method for feature extraction, the USPS database is used in the experiments [116]. This database consists of USA city names, street names and numerals, where the images of city names and street names are gray scale and the images of numerals are binary. The images in the database are written by different people with varied writing styles, such as printed, cursive and mixed writing styles. Further, the paper and writing tools are not restricted. As a consequence, the quality of images varies considerably.

B: Results of Skeletonization for Gray-Scale Images Using Gabor Filter

Extracting features directly from gray-scale images reduces information loss and distortions caused by binarization. In the following, we show some experimental results.

HAND PRINTED:
Figure 2.21(a) shows a slant and tilt-corrected gray-scale image of good quality. The writing style of the word in Figure 2.21(a) is hand printed. The orientation strength functions at optimal directions are displayed in Figure 2.21(b), where the high intensity at (x_i, y_i) represents high value of $f_m(x_i, y_i)$. Figure 2.21(c) shows the result after initial skeletonization, where the initial skeleton is not perfect and there are redundant data. Figure 2.21(d) shows the final result of our method.

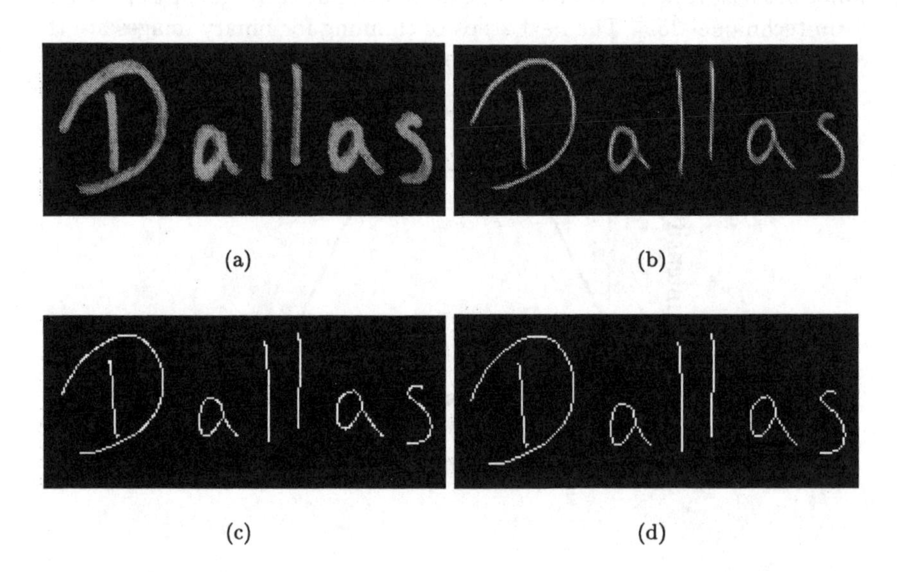

FIGURE 2.21. Feature extraction from a hand printed image.

CURSIVE SCRIPT:
Figures 2.22(a) and (c) show the slant and tilt-corrected cursive script images. The extracted skeletons of the images are shown in Figures 2.22(b) and (d), respectively. As we can see from the figures, the skeletons are very smooth.

NOISY IMAGES:
Figure 2.23(a) shows an original handwritten word image and Figure 2.23(b) shows its slant and tilt-corrected version. Extracting features from such images is a challenging task as there are several thin noisy lines caused by the writing tool. Conventional methods will produce many noisy branches,

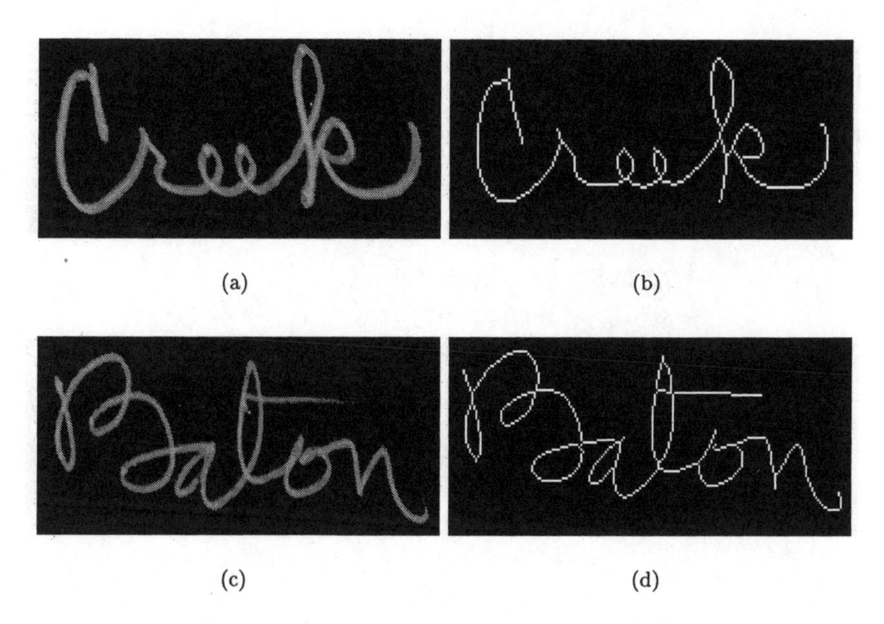

(a) (b)

(c) (d)

FIGURE 2.22. Feature extraction from cursive writings.

small holes, and significant distortions. For the human, however, it is not difficult to identify the line-segments caused by noise. Our method is quite successful in dealing with such low quality images. Figure 2.23(c) shows the orientation strength functions at optimal directions, where the noisy lines caused by the writing tool have been erased. This property is attributed to the Gabor filter whose parameters have been automatically adjusted using the methods presented in the previous sections to adapt to the average line width. We also find that the surface of the orientation strength functions is much smoother than that of the intensity of the original or slant and tilt-corrected image. Furthermore, the line width after deleting edge points is much smaller than that in the original image. The above properties are desirable for robust feature extraction. Figure 2.23(d) shows the result of this example. In this result, there are one broken line at an X-cross and few short noisy branches, which can be easily removed by post-processing [220].

For comparison, Figure 2.24 shows some feature extraction results for gray-scale images using Lee and Kim's method 2 [156]. The images in the left column are results of Lee and Kim's method 2 and the images in the right column are results after deleting redundant data. Figures 2.24(a) to (f) are obtained from good quality gray-scale images by Lee and Kim's method. However, their method produces the skeletons with lots of noisy branches and short lines. Their method is more likely to break lines, specially, at crosses. Figure 2.24(g) and (h) are examples produced by Lee and Kim's method for noisy image. This shows that their method cannot deal with

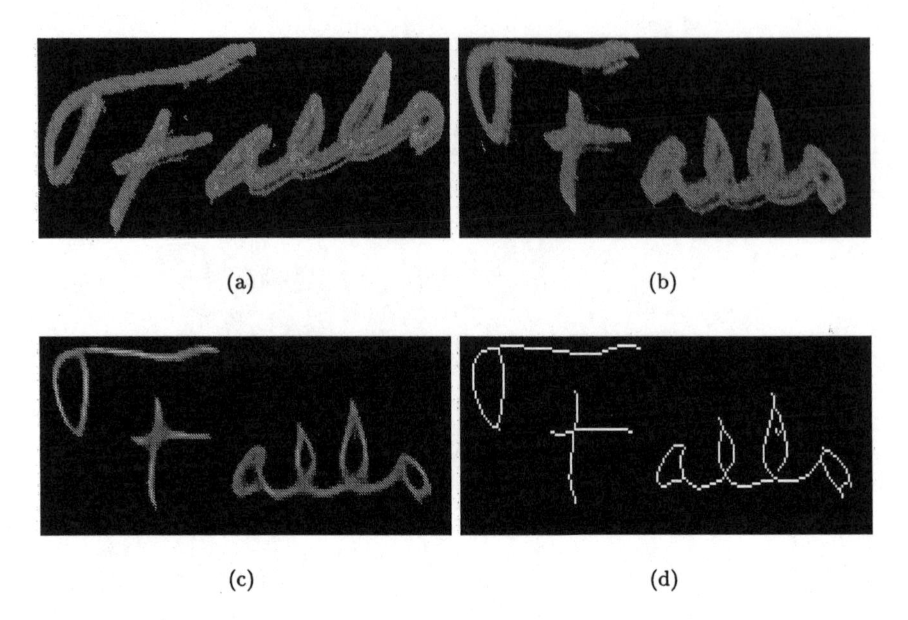

(a) (b)

(c) (d)

FIGURE 2.23. Feature extraction from a noisy image.

the problems caused by noisy thin line segments. When compared with Figure 2.23(d), our method handles these problems remarkably well.

C: Results of Skeletonization for Binary Images Using Gabor Filter

Techniques for extracting features from binarized images still dominate the literature and commercial systems. In the following, we will show some experimental results for binary images.

BINARY IMAGES OF GOOD QUALITY:
Figure 2.25(a) shows a pseudo gray-scale image that is converted from a binary image. Figure 2.25(b) gives the orientation strength functions at optimal directions. Note that the image of the orientation strength functions is very smooth and there is no flat region. All pixels in this image are represented in terms of peak points, ridge points, and edge points. Figures 2.25(c) and (d) are the initial skeleton and the final result of feature extraction, respectively. The final skeleton of the image is smooth and there is no significant distortion at the cross. Figure 2.26 shows more results of our method. For comparison, we show some examples of the method proposed in [5] for binary images in Figure 2.27.

BINARY IMAGE WITH A FILLED HOLE:
This example poses a significant challenge. Figure 2.28(a) shows the pseudo gray-scale image which is converted from a binary image shown in Figure 2.11. In Figure 2.28(b), we show the orientation strength functions

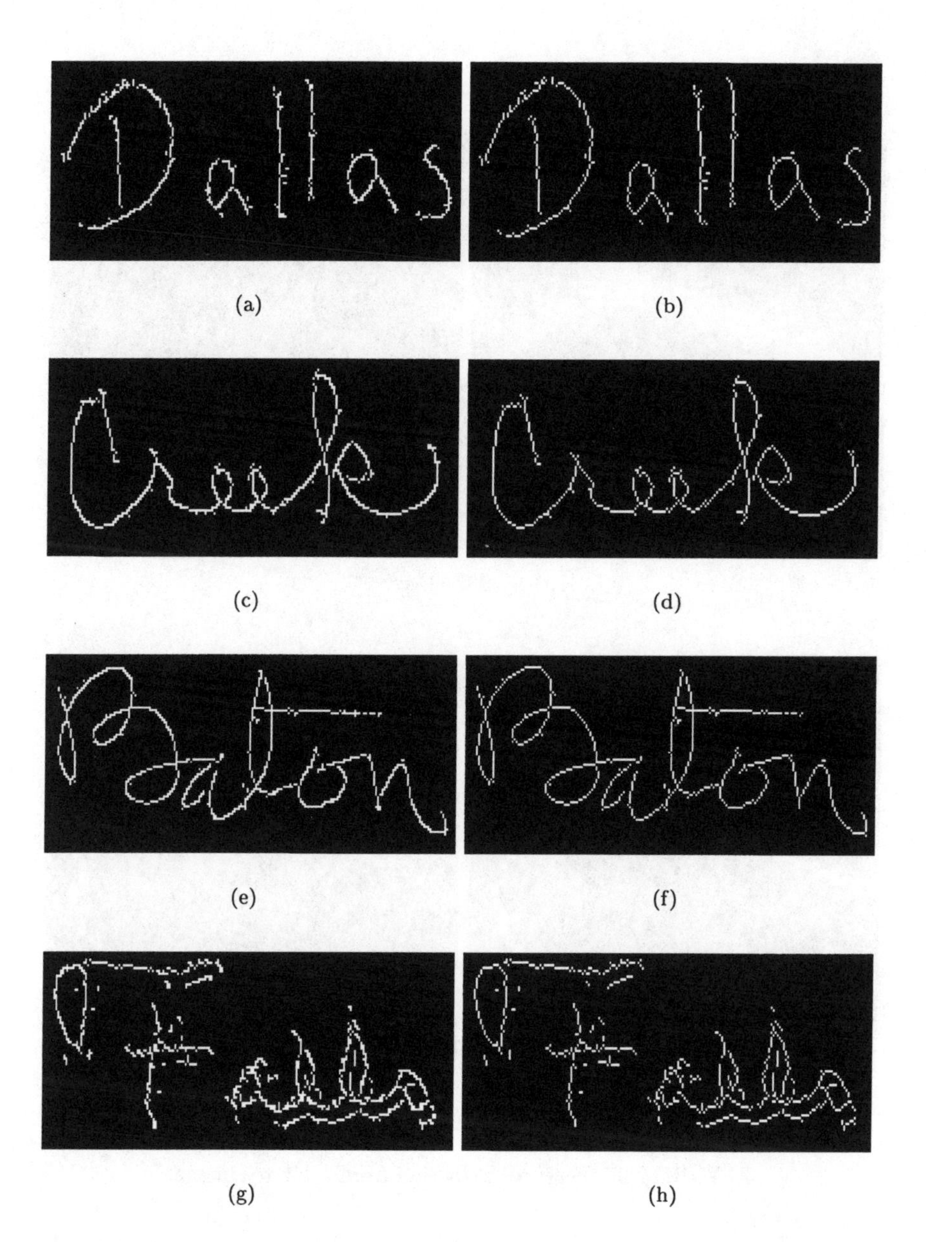

FIGURE 2.24. Results of Lee and Kim's method for gray-scale images.

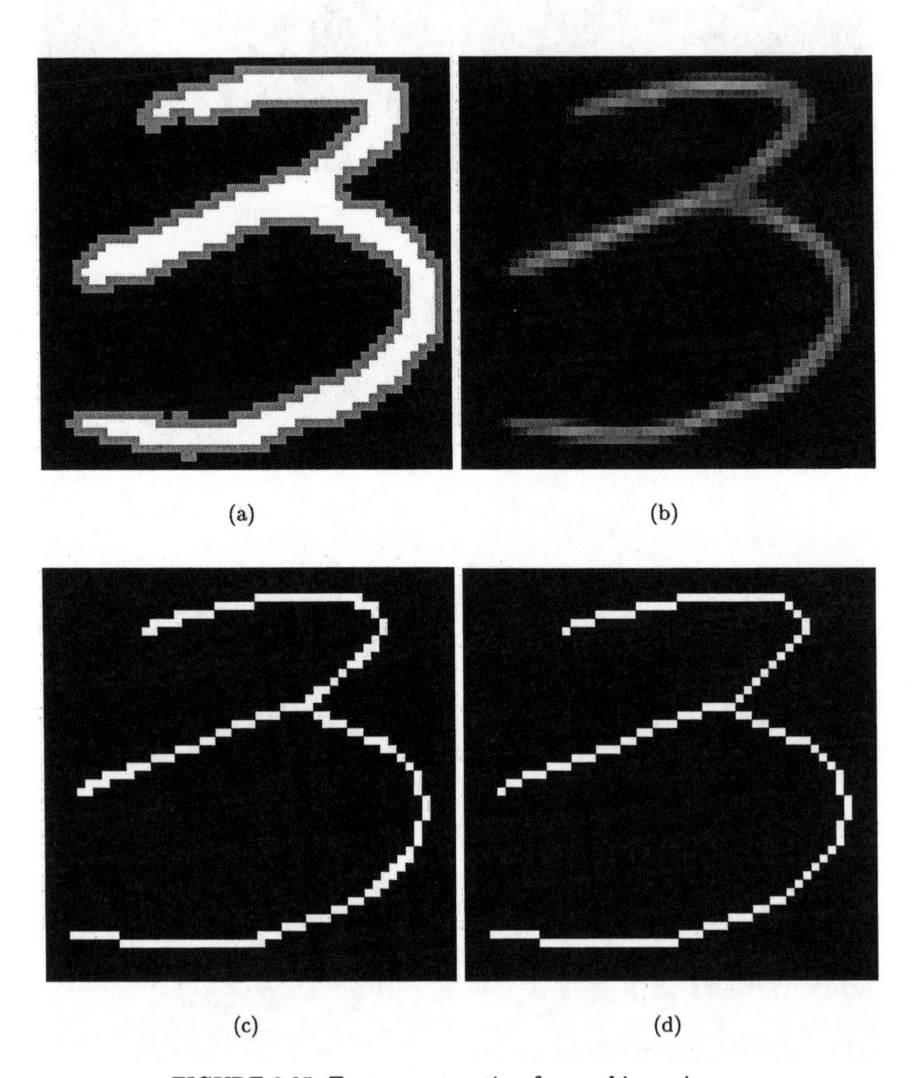

FIGURE 2.25. Feature extraction from a binary image.

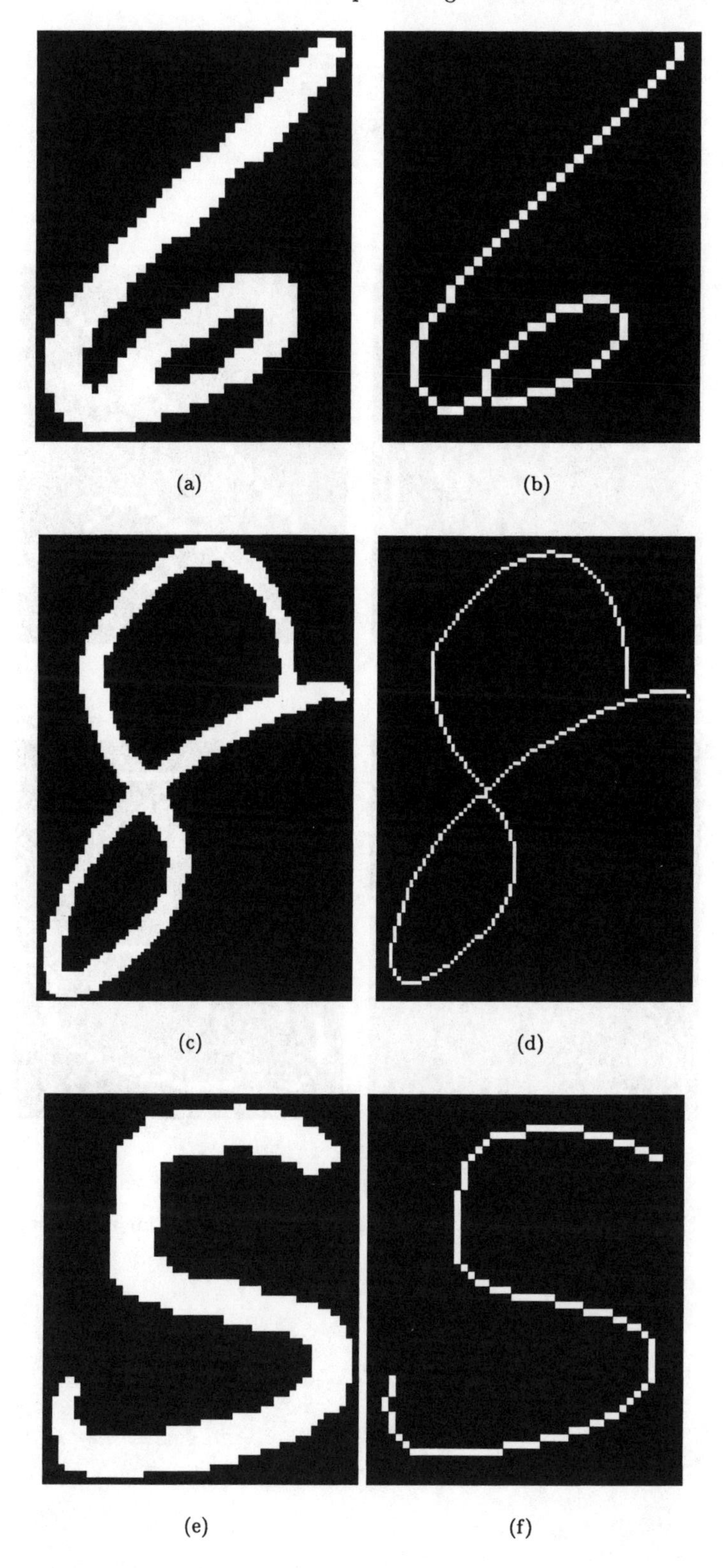

(a) (b)

(c) (d)

(e) (f)

FIGURE 2.26. Examples of skeletonization from binary images.

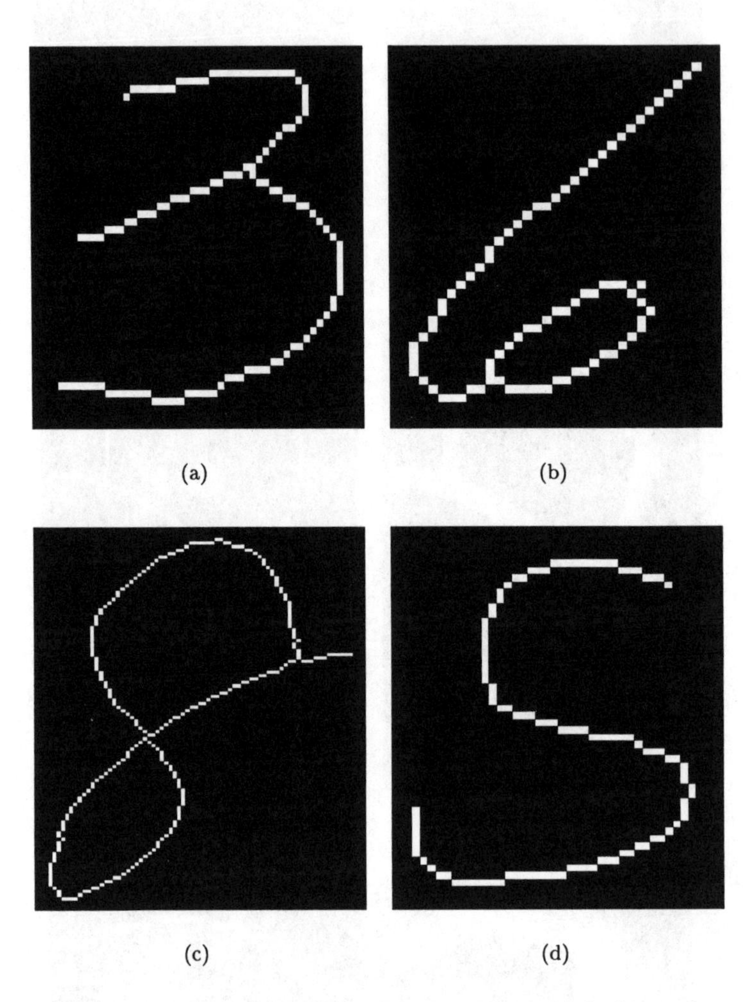

FIGURE 2.27. Results of Arcelli and Baja's method [5] for binary images.

at optimal directions. Figures 2.28(c) and (d) are the initial skeleton and the final result of feature extraction, respectively. Our method uses the line width to convert the binary image into a pseudo gray image and successfully extracts the skeleton without introducing significant distortions. Although this method does not preserve the connectivity perfectly, it is still able to produce good results from images with filled holes. For such cases, existing algorithms may fail to deal with holes filled by thick line segments. For instance, Arcelli and Baja's method can obtain good results from good quality binary images, but fails in this case. Figure 2.29 gives the result obtained by their method from the image shown in Figure 2.11.

2.3.3 Extracting Oriented Segments Using Gabor Filters

For handwriting recognition, the information about the presence of line segments at given orientations can be used to represent prototypes of handwritten patterns. Therefore, local detection of line segments seems to be an adequate feature extraction method [154]. Many filters and feature detectors can fulfill the task of extracting oriented segments from images. In [154], Lee extracted features using 3×3 Kirsch edge detector, which may produce two line segments from one thick line. Since 2-D Gabor filters can adapt the width of lines, achieve the theoretical lower bound of joint uncertainty in space and frequency domains, and are robust to noise, we use them to extract oriented segments. The oriented segments are obtained by the Gabor filters and divided into eight groups according to their orientations [38]. The oriented segments are extracted by the Gabor filters; Figure 2.30 shows some examples.

2.4 Concluding Remarks

In this chapter, we have presented major pre-processing techniques in handwriting recognition, which are used to reduce noise and variations caused by different writing styles and the imaging process. We have introduced methods for feature extraction from binary images. We have also presented a powerful method using Gabor filters for handwritten image skeletonization and demonstrated that this method is robust to noise. The quality of skeletons produce by this method is excellent. In addition, every step of this method can be implemented in a fully parallel fashion, because the computation involved in each step is based on local information. Refining skeletons can be done by a fully parallel algorithm introduced in [260]. This method requires more computation than conventional methods due to Gabor filtering operations; however, these filtering operations are amenable to being implemented by optoelectronic processors at high speed [246]. One of the most important advantages of this method is that it is effective to

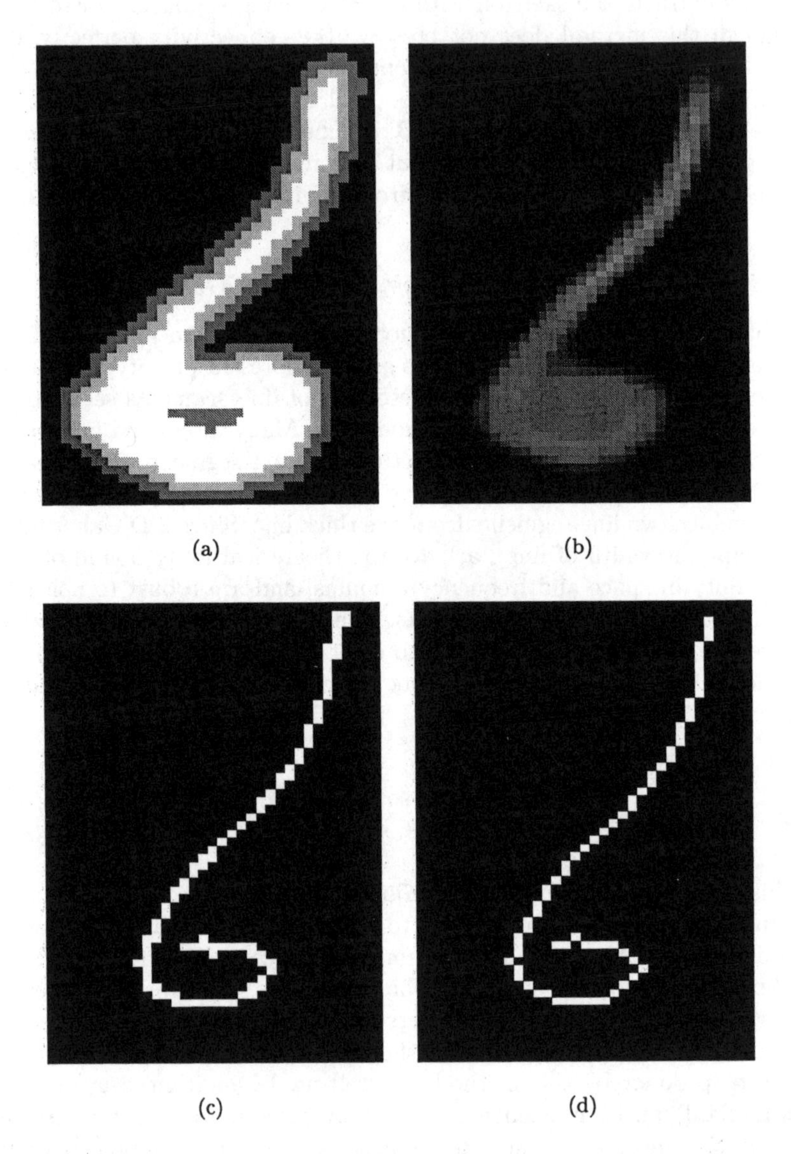

FIGURE 2.28. Skeletonization from binary image with a filled hole.

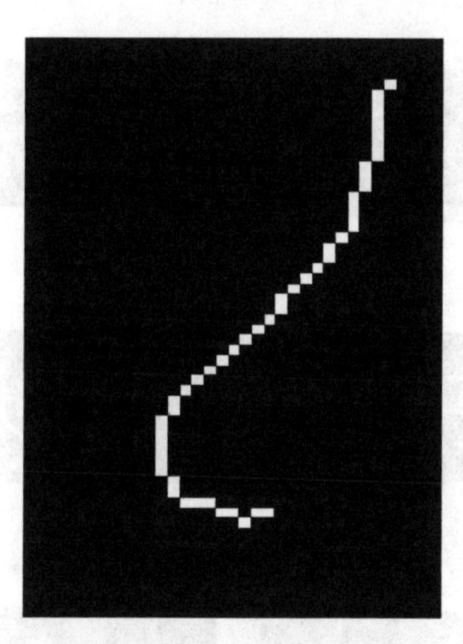

FIGURE 2.29. The result of Arcelli and Baja's method [5] for binary image with a filled hole.

both gray-scale images and binary images.

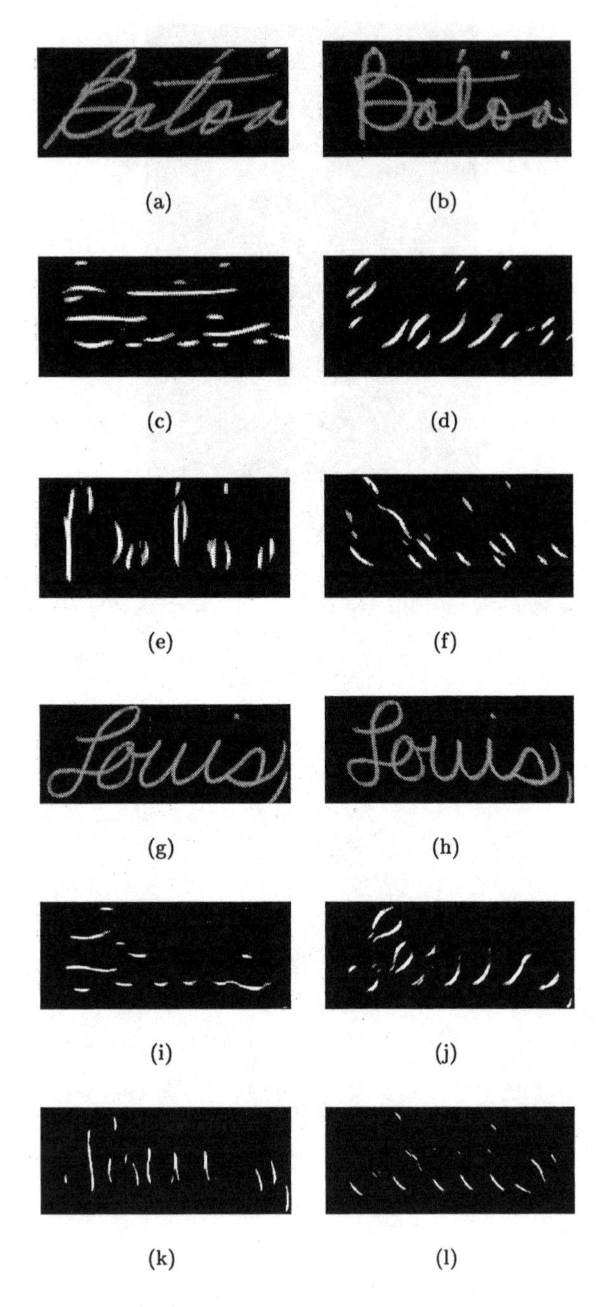

FIGURE 2.30. The output energy of Gabor filters for the given words. (a) and (g) The original images. (b) and (h) The slant and tilt corrected images. (c), (d), (e), (f), (i), (j), (k) and (l) The examples of the Gabor filter output. The angles of Gabor filters are 0° in (c) and (i), 45° in (d) and (j), 90° in (e) and (k) and 135° in (f) and (l).

3

Hidden Markov Model-Based Method for Recognizing Handwritten Digits

In this chapter, we introduce the theory of HMMs and present an HMM-based method for recognizing unconstrained handwritten numerals.

The following section introduces the basic concepts of discrete-time Markov processes and HMMs. We discuss various types of HMMs and parameter re-estimation and evaluation techniques. In Section 3.2, we present a new approach that uses the statistical and structural information for handwritten numeral recognition. In this approach, we use the HMM to model the statistical information. The states of HMMs are used to model the statistics of short line-segments. For modeling structural information, we introduce the concept of *macro-states*. This method is evaluated over a standard handwritten digit database (CEDAR CDROM1), the experimental results are presented in Section 3.3.

3.1 Theory of Hidden Markov Models

3.1.1 Markov Process

Let us take weather forecasting as an example. We may label the weather as one of four categories: sunny, cloudy, rainy, or snowy, each of which is a *state*. If today is a rainy day, tomorrow will more likely be a rainy or cloudy day than a sunny day. Therefore, the weather of the following day depends on the weather of today to a certain degree. There are some constraints between the weather of two continuous days. Such a process can be described by a Markov process, where the weather is represented by one of the four states on a particular day (time).

Let S be a countable set. Each $i \in S$ is called a state and S is called the *state-space*. Let q_t be a random variable with value in S and $\mathbf{A} = (a_{ij} | i, j \in S)$ denote a matrix in which a_{ij} is the probability of state transition from i to j. Consider a simple system which may be described at any time as being in state $q_t \in S$ as shown in Figure 3.1.

This system has four states and is fully connected as any state can be reached from any other state including itself. Each directed line is a transition from one state to another state whose probability is indicated by the number alongside the line. State transition takes place only at dis-

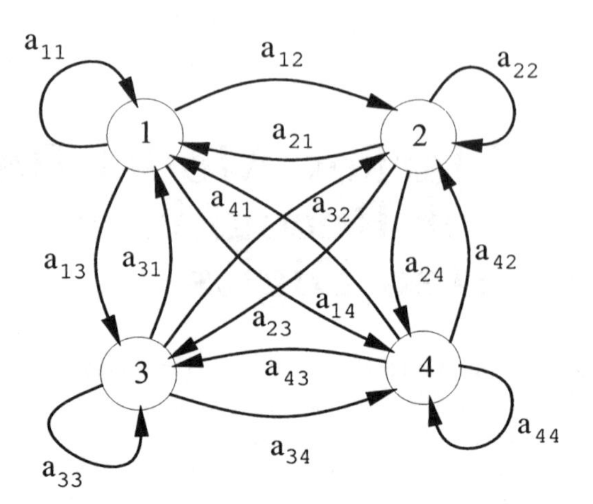

FIGURE 3.1. A Markov chain with 4 states.

crete times, $\{1, \cdots, t, t+1, \cdots, T\}$. Theoretically, all previous states can influence the current state in such a system. However, in real-world applications, it can be computationally intractable and unnecessary, because in reality states that are far apart usually have little influence on each other. For instance, yesterday's storm may result in heavy cloudy day today, but it will not likely to have a significant influence on the weather condition the same day next month. For most systems we assume that the process $(q_t | 1 \le t \le T)$ is a discrete-time, first-order Markov chain [185]. This means that the process $(q_t | 1 \le t \le T)$ must satisfy the following two conditions:

(i) $Pr(q_1 = i_1) = \pi_{i_1}, \qquad 0 \le \pi_{i_0} \le 1;$

(ii) $Pr(q_t = j | q_1 = i_1, \cdots, q_{t-1} = i) = Pr(q_{t+1} = j | q_t = i),\ t \ge 0,$

where $(\pi_{i_1} | i_1 \in S)$ are called the initial state probabilities and $\sum_{i_1} \pi_{i_1} = 1$. In this way, the probabilistic description is truncated to just the current and the previous state. Usually, it is sufficient to consider only a time-invariant system with the following state transition probabilities:

$$a_{ij} = p(q_t = j | q_{t-1} = i), \qquad i, j \in S. \tag{3.1}$$

Therefore, the system shown in Figure 3.1 at time t can be characterized by a single state and has a time-invariant matrix:

$$\mathbf{A} = \begin{pmatrix} a_{11} & a_{12} & a_{13} & a_{14} \\ a_{21} & a_{22} & a_{23} & a_{24} \\ a_{31} & a_{32} & a_{33} & a_{34} \\ a_{41} & a_{42} & a_{43} & a_{44} \end{pmatrix}, \tag{3.2}$$

where state transition probabilities a_{ij} obey the following standard stochastic constraint:

$$\sum_j a_{ij} = 1. \tag{3.3}$$

The above model is an observable Markov model, because each state of the model corresponds to a physical event that can be observed. However, since in many applications not all states are *directly* observable, this model is too restrictive to be applicable to many problems of interest [111].

3.1.2 Hidden Markov Models

For handwriting recognition, discrete-time features are used in both on-line and off-line cases, where *time* can be real time or an index of feature frames, etc. For instance, the motion of pen with respect to a tablet is sampled to generate the discrete-time features derived from the pen motion information [200]. Elms extracted discrete-time feature vectors sequentially from pixels line by line (column or row) [80]. In this case, the time represents the index of columns or rows. Let the features of a system be a random sequence, $O_t \in \mathbf{O}$, in which a short segment can be considered statistically stationary [71] and be modeled by the Hidden Markov model (HMM) [111].

Hidden Markov models have been widely used in speech recognition [209, 199], handwriting recognition [58, 80] and automatic control [79]. In the following we give a brief discussion of the concept of HMM. Let us consider the urn-and-ball experiment [209]. As illustrated in Figure 3.2, we assume there are N urns in a room. Each urn contains a large number of colored balls. There are M distinct colors of the balls. A person is performing an experiment behind a curtain. This person chooses an urn according to a random process. He chooses a ball from the selected urn at random and shows us the ball. We record the color of the ball as our observation. The ball is then replaced in the urn from which it was selected. A new urn is then chosen according to the random selection process and the ball selection process is repeated. After many repeats, the entire process generates a finite observation sequence of colors, which we would like to model as the observable output of a Markov process. As only the sequence of colored balls can be observed in front of the curtain, whereas the urns from which the balls in the sequence were taken is *hidden*, this Markov model is called hidden Markov model. In this simplest HMM, each state corresponds to a particular urn. The choice of urns is determined by a state transition matrix and the probabilities of colors of the picked balls are defined for each state (urn).

The above example clearly illustrates the concept of HMM. In general, the HMM has the following elements:

- N is the number of states in the model. Although the states are hidden or the state at t is unknown, the number of states is pre-set for many practical applications. The set of states in the model is $S = \{1, 2, \cdots, N\}$, the state at time (index) t is denoted as q_t.
- T is the length of the observation sequence, $\mathbf{O} = \{O_1, O_2, \cdot, O_T\}$, where O_t is the observation at t (in this example, O_t is the color

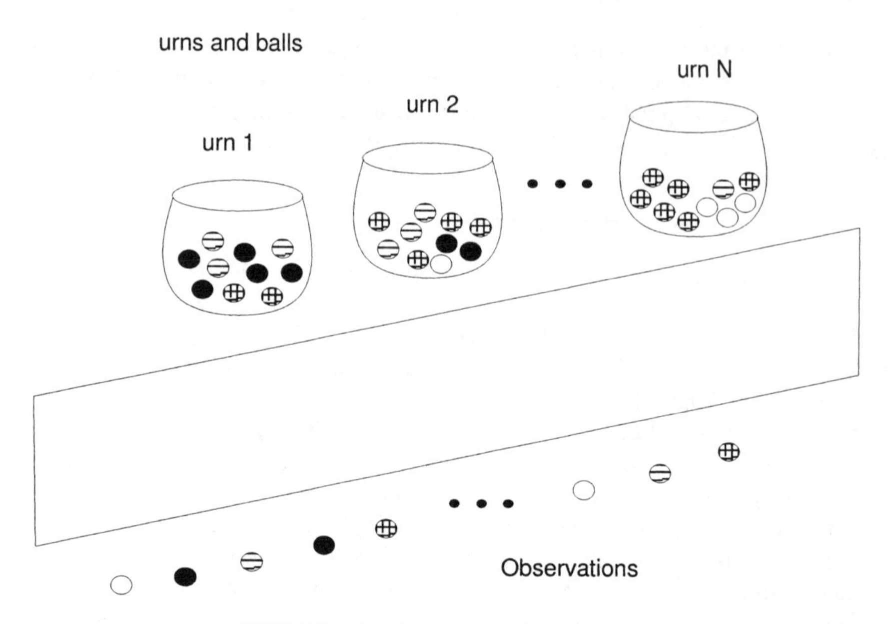

FIGURE 3.2. The urn and ball example

of the ball selected at t). The observation sequence results in a set of states, $\mathbf{X} = \{q_1, q_2, \cdots, q_t, \cdots, q_T\}$. It should be noted that the state set $\mathbf{X}$ is different from the state set S. An HMM having a set of N states S produces a sequence of observation in the corresponding state sequence $\mathbf{X}$, where $q_t \in S$.

- M is the number of distinct observation symbols in the model. The observation symbols correspond to the physical output of the system. We denote the individual symbols as $\mathbf{V} = \{v_1, v_2, \cdots, v_M\}$. Note that the discrete symbols are only used in discrete HMMs (DHMMs).

- $S = \{1, 2, \cdots, N\}$ is a set of states. Each state is considered to possess some measurable, distinctive properties of events [111].

- $\mathbf{A} = \{a_{ij}\}$ is the state transition probability matrix, where a_{ij} denotes the state transition probability from state i to state j. Note that $\mathbf{A}$, the state transition probability matrix, is independent of time.

- $\pi = \{\pi_i\}$ is the initial state distribution, where $\pi_i = Pr(q_1 = i)$.

- $\mathbf{B} = \{b_j(O_t)\}$ is the observation symbol probability distribution, where $b_j(O_t) = Pr(O_t | q_t = j)$. In the HMM, this probability distribution of the observation at any time is determined only by the current state of the Markov chain. In the case of DHMMs, $b_j(O_t)$ are the discrete probability distributions for state j, where $O_t \in \mathbf{V}$. Figure 3.3

shows the discrete probability distributions for the urn-and-ball example. In the continuous HMM, $b_j(O_t)$ is the output probability for state j and usually assumed to be a Gaussian function or a Gaussian mixture. In some systems, $b_j(O_t)$ can be a hybrid of discrete and continuous functions.

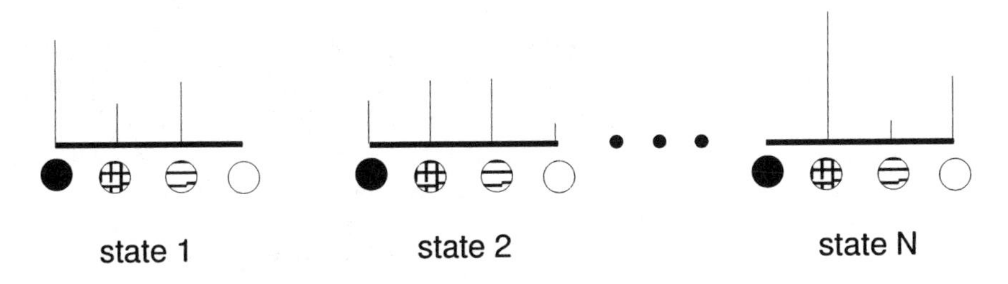

FIGURE 3.3. Example of discrete probability distributions

In above basic HMM, we need to specify values for (N, M) and to estimate three probability distributions $(\mathbf{A}, \mathbf{B}, \pi)$ from training data. For convenience, we can express the basic HMM as follows:

$$\lambda = (\mathbf{A}, \mathbf{B}, \pi). \tag{3.4}$$

3.1.3 Basic Algorithms for HMMs

Given the definition of HMM, there are three key problems that must be solved when we use the HMM in a real-world process:

(1) Evaluation: Given an observation sequence, $\mathbf{O} = \{O_1, O_2, \cdots, O_T\}$, and a model $\lambda = (\mathbf{A}, \mathbf{B}, \pi)$, what is the probability of the model that has produced the observed sequence? And how to efficiently compute the probability? In other words, evaluation enables us to choose the model that best matches the observation sequence.

(2) Decoding: Given an observation sequence, $\mathbf{O} = \{O_1, O_2, \cdots, O_T\}$, and a model, $\lambda = (\mathbf{A}, \mathbf{B}, \pi)$, how to choose the optimal state sequence that results in the highest probability for the model to produce the observation sequence? Or put simply, how to uncover the *hidden* part of the model? It is particularly useful in analyzing the structure of the model. This makes it possible to combine statistical and structural information for achieving good recognition performance [36].

(3) Re-estimation: How do we adjust the model parameters to maximize the score in problem (1)? Re-estimating the parameters of a model is to create the best model from its initial guess to reflect the nature of the real-world process.

Evaluation: Forward-Backward Algorithm

In statistical or Bayesian paradigms, decision-making is based on the concept of the maximum *a posteriori* (MAP) probability:

$$Pr(w|\mathbf{O}, \lambda_w) = Pr(w)\frac{Pr(\mathbf{O}|\lambda_w)}{Pr(\mathbf{O})}, \tag{3.5}$$

where $Pr(w)$ is the *priori* probability that the given pattern belongs to the class w, λ_w is the model for class w, $Pr(\mathbf{O}|\lambda_w)$ is the conditional probability of the observation set $\mathbf{O}$ for the given model λ_w, $Pr(\mathbf{O})$ is the probability of the observation sequence, and $Pr(w|\mathbf{O}, \lambda_w)$ is the probability that the pattern belongs to class w for the observation set and model. This is the *a posteriori* probability. Because $Pr(\mathbf{O})$ is not related to class w, it is irrelevant to recognition. Therefore, (3.5) can be rewritten as follows:

$$Pr(w|\mathbf{O}, \lambda_w) \approx Pr(w)Pr(\mathbf{O}|\lambda_w). \tag{3.6}$$

We can make a classification decision by

$$w^* = \arg\max_w \{Pr(w)Pr(\mathbf{O}|\lambda_w)\}, \tag{3.7}$$

where w^* is the estimated pattern class. If $Pr(w)$ is unknown, it is set to a constant and the MAP estimation becomes a maximum likelihood (ML) estimation. As a result, the problem of recognition in (3.6) becomes that of estimating $Pr(\mathbf{O}|\lambda_w)$. Let us introduce the concept of state into the MAP estimation. We express $Pr(\mathbf{O}|\lambda_w)$ as follows:

$$Pr(\mathbf{O}|\lambda_w) = \sum_{\text{all } \mathbf{X}} Pr(\mathbf{O}, \mathbf{X}|\lambda_w), \tag{3.8}$$

where

$$Pr(\mathbf{O}, \mathbf{X}|\lambda_w) = Pr(\mathbf{O}|\mathbf{X}, \lambda_w)Pr(\mathbf{X}|\lambda_w). \tag{3.9}$$

For every fixed state sequence $\mathbf{X} = \{q_1, q_2, \cdots, q_T\}$, the probability of the observation sequence $\mathbf{O}$, $Pr(\mathbf{O}|\mathbf{X}, \lambda_w)$ is

$$Pr(\mathbf{O}|\mathbf{X}, \lambda_w) = b_{q_1}(O_1)b_{q_2}(O_2)\dots b_{q_T}(O_T). \tag{3.10}$$

Due to the assumption of the first-order Markov chain, the probability of such a state sequence $\mathbf{X}$ is

$$Pr(\mathbf{X}|\lambda_w) = \pi_{q_1} a_{q_1 q_2} a_{q_2 q_3} \cdots a_{q_{T-1} q_T}. \tag{3.11}$$

Now, the probability of the observation sequence $\mathbf{O}$ for λ_w can be expressed as follows:

$$Pr(\mathbf{O}|\lambda_w) = \sum_{\text{all } \mathbf{X}} \prod_{t=1}^{T} a_{q_{t-1} q_t} b_{q_t}(O_t), \tag{3.12}$$

where $a_{q_0 q_1}$ denotes π_{q_1} for simplicity.

Unfortunately, the complexity of directly computing the above equation is in the order of $2TN^T$, since there are N^T possible state sequences. This is unfeasible for even small values of N and T [209]. Clearly, direct computation of (3.12) is prohibitive in real-world applications for which we must find a more efficient solution.

The Forward-Backward algorithm has been proposed as a solution to this problem [209, 111]. In this algorithm, there are two key variables: forward variable $\alpha_t(i)$ and backward variable $\beta_t(i)$. A forward variable is defined as follows:

$$\alpha_t(i) = Pr(O_1, O_2, \cdots, O_t, q_t = i|\lambda_w). \qquad (3.13)$$

$\alpha_t(i)$ actually is the probability of the partial observation sequence, $\{O_1, O_2, \cdots, O_t\}$ up to t and the state is i at time t, given λ_w. This probability can be calculated inductively:

1 Initialization:

$$\alpha_1(i) = \pi_i b_i(O_1), \qquad 1 \leq i \leq N. \qquad (3.14)$$

Step 1 initializes the forward probabilities as the joint probability of state $q_1 = i$ and the observation O_1 for all possible states.

2 Induction:

$$\begin{aligned} \alpha_t(j) &= \left[\sum_{i=1}^{N} \alpha_{t-1}(i) a_{ij}\right] b_j(O_t), \\ & 2 \leq t \leq T \text{ and } 1 \leq j \leq N. \end{aligned} \qquad (3.15)$$

In this step, the joint probability of state j at t and the partial observation $\{O_1, \cdots, O_t\}$ is the output probability $b_j(O-t)$ multiplied by the summation of the product of the joint probabilities of all possible states at $t-1$ and the state transition probabilities to the next state j. The procedure for a single state is illustrated in Figure 3.4(a). At $t+1$, $\alpha_{t+1}(j)$ can be calculated in the same way. Figure 3.4(b) depicts the complete procedure for computing forward variables.

3 Final output probability:
The probability of $Pr(\mathbf{O}|\lambda_w)$ is given by

$$Pr(\mathbf{O}|\lambda_w) = \sum_{i \in S_F} \alpha_T(i), \qquad (3.16)$$

where S_F is a set of possible final states.

The computation of forward variables leads to a lattice structure shown in Figure 3.4(b), in which the transition from any state i to state j at t is allowed. Since there are only N states (nodes) at each column in the lattice, we can calculate the joint probabilities of a partial observation and

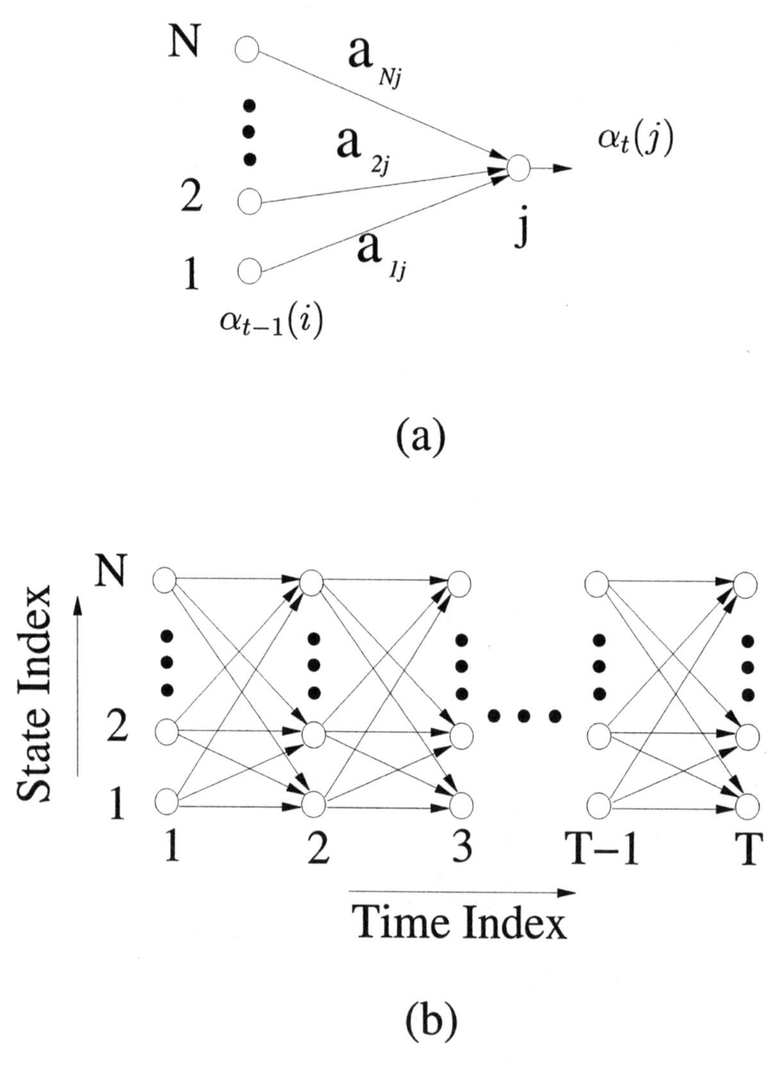

FIGURE 3.4. The inductive computation of forward variables $\alpha_t(j)$. (a) Single node operation of forward variable computation. (b) Implementation of the computation of forward variable in terms of a lattice of observations and states for a fully-connected HMM.

the states by columns ordered in time indexes. Consequently, the computation of $Pr(\mathbf{O}|\lambda_w)$ is reduced to the order of N^2T. Compared to the direct computation, the computation of forward variables in the lattice structure is far more efficient.

In a similar way, we define a backward variable $\beta_t(i)$ as follows:

$$\beta_t(i) = Pr(O_{t+1}, O_{t+2}, \cdots, O_T | q_t = i, \lambda_w), \tag{3.17}$$

where $\beta_t(i)$ is the probability of the partial observation sequence from $t+1$ to the end T, given state i at t and λ_w. Similarly, we can also calculate the backward variable in an inductive manner:

1. Initialization:

$$\beta_T(i) = 1, \qquad 1 \leq i \leq N. \tag{3.18}$$

2. Induction:

$$\beta_t(i) = \sum_{j=1}^{N} a_{ij} b_j(O_{t+1}) \beta_{t+1}(j), \tag{3.19}$$

where $t = T-1, T-2, \cdots, t, \cdot, 1$ and $1 \leq i \leq N$.

3. Final output probability:

$$Pr(\mathbf{O}|\lambda_w) = \sum_{\text{all } \mathbf{x}} \pi_i b_i(O_1) \beta_1(i). \tag{3.20}$$

In step 2, to stay in state i and to account for the partial observation sequence from $t+1$ onwards, every possible state j at $t+1$ must be taken into consideration. The state $q_{t+1} = j$ accounts for the transition from state i to j and the observation O_{t+1} in state j. The sequence of operations required for the calculation of $\beta_t(i)$ is illustrated in Figure 3.5. The whole process for computing backward variables is very similar to that for $\alpha_t(i)$, but in the inverse order of time.

Step 3 gives the final output probability as the sum of the product of every backward variable at $t = 1$, the corresponding initial state probability, and the corresponding output probability of the observation O_1. Note that the backward variables are not necessary in computing $Pr(\mathbf{O}|\lambda_w)$. They are introduced here as they will be used in the training algorithms of HMMs.

State Decoding: the Viterbi Algorithm

The forward-backward algorithm can be used to solve the evaluation problem; however, it does not provide the information about the hidden part of the HMM, i.e., the state sequence. Although the state sequence of the HMM cannot be uncovered, it can be interpreted in some meaningful way. An important use of the recovered state sequence is to retrieve the behavior of individual states and the state structure of the model. The Viterbi

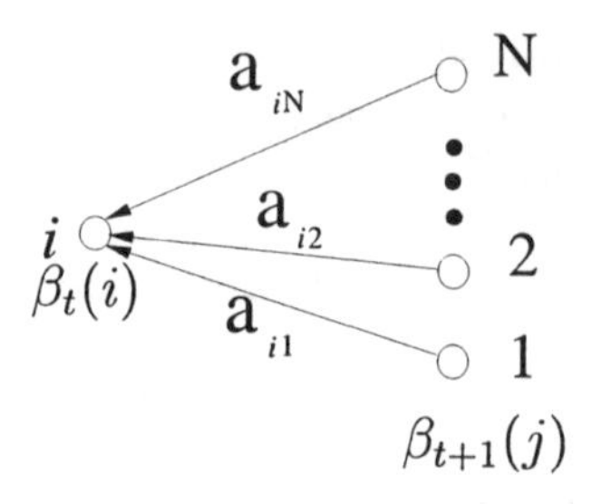

FIGURE 3.5. Single node operation of backward variable computation.

algorithm is an efficient algorithm to find an optimal solution [249]. For the HMM, the Viterbi algorithm is used to find the optimal state path that has the highest probability associated with the given observation sequence. With the optimal state path, we can learn the relationships between states. Such state relationships are very useful in identifying patterns with similar statistics but different state structure. In order to find the single best state sequence, the optimality criterion is given by

$$\delta_t(i) = \max_{q_1, q_2, \cdots, q_{t-1}} Pr(q_1, q_2, \cdots, q_{t-1}, q_t = i, O_1, O_2, \cdots, O_t | \lambda_w). \quad (3.21)$$

$\delta_t(i)$ is the highest probability along a single path, which accounts for the partial observation sequence $O_1, O_2, \cdots, O_t$ and ends at state i. If the state at time $t+1$ is $q_{t+1} = j$, the highest probability to generate the partial observation can be obtained from the following operation:

$$\delta_{t+1}(j) = [\max_i \delta_t(i) a_{ij}] \cdot b_j(O_{t+1}). \quad (3.22)$$

In this way, we can retrieve the optimal state path. A formal description of the Viterbi algorithm used in a single hidden Markov model can be summarized as follows:

Step 1: Initialization.

$\alpha_1(i) = \pi_i b_i(O_1),$

$\Phi_1(i) = 0 \qquad 0 \le i \le N-1.$

Step 2: Recursion.

$\alpha_t(j) = \max_i [\alpha_{t-1}(i) a_{ij} b_j(O_t)],$

$\Phi_t(j) = \arg\max_i [\alpha_{t-1}(i) a_{ij}] \qquad 2 \le t \le T.$

Step 3: Termination. (S_F is the final state set.)

$p^*(\mathbf{O}|\lambda_w) = \max_{s \in S_F} [\alpha_T(s)],$

$q_T = \arg\max_{s \in S_F} [\alpha_T(s)].$

Step 4: State path backtracking.

$q_t = \Phi_{t+1}(q_{t+1}).$

The Viterbi algorithm can solve the evaluation problem as well as decoding problem. As discussed above, the Viterbi algorithm obtains the maximum of $Pr(\mathbf{O}, \mathbf{X}|\lambda_w)$ over all possible state sequence $\mathbf{X}$, while the forward-backward algorithm finds the true output probability $Pr(\mathbf{O}|\lambda_w)$ that is the summation of $Pr(\mathbf{O}, \mathbf{X}|\lambda_w)$ over all $\mathbf{X}$. Disregarding the difference between $Pr(\mathbf{O}|\lambda_w)$ and the maximum of $Pr(\mathbf{O}, \mathbf{X}|\lambda_w)$, the probabilities obtained from the forward and the Viterbi algorithms are very similar [207]. Since the Viterbi algorithm can operate using only additions in the logarithmic domain and, at the same time, can obtain the optimal state sequence, it has been widely used in many systems [34], particularly in speech recognition.

Re-Estimation Algorithms

The most important issue of HMMs is how to train a particular HMM λ_w to correctly represent the observation sequences from training samples from class w. As there is no analytical methods existing for parameter optimization to maximize the probability of the observation, iterative procedures are used to solve the problem of re-estimation of HMMs.

A: Baum-Welch Re-estimation Algorithm

Baum and his colleagues proposed the forward-backward algorithm for the re-estimation of HMMs [10, 11] known also as the Baum-Welch algorithm. This algorithm uses the same optimization technique as the expectation-modification (EM) algorithm [72] to improve the model iteratively. Usually, the surface of the output probability is complex and has many local maximums, the Baum-Welch algorithm can guarantee to reach a local maximum. If the initialization values of model parameters are carefully chosen, this local maximum may be the global maximum or at least near the global maximum.

To formally describe the Baum-Welch algorithm, we define $\zeta_t(i,j)$, as follows:

$$\begin{aligned}
\zeta_t(i,j) &= Pr(q_t = i, q_{t+1} = j|\mathbf{O}, \lambda_w) \\
&= \frac{\alpha_t(i)a_{ij}b_j(O_{t+1})\beta_{t+1}(j)}{Pr(\mathbf{O}|\lambda_w)} \\
&= \frac{\alpha_t(i)a_{ij}b_j(O_{t+1})\beta_{t+1}(j)}{\sum_{k=1}^{N}\sum_{l=1}^{N}\alpha_t(k)a_{kl}b_l(O_{t+1})\beta_{t+1}(l)}.
\end{aligned} \tag{3.23}$$

We can see that $\zeta_t(i,j)$ is the probability of being in state i at time t and state j at time $t+1$ for the given model and the observation sequence. We define $\gamma_t(i)$ as the probability of being in state i at t for the given model and the observation sequence. This probability can be expressed as the summation of $\zeta_t(i,j)$:

$$\gamma_t(i) = \sum_{j=1}^{N} \zeta_t(i,j) \qquad 1 \leq t \leq T-1. \tag{3.24}$$

According to the above, we can interpret the summation of $\zeta_t(i,j)$ over the time index t as the expected number of transitions from state i to j. Similarly, the summation of $\gamma_t(i)$ over t can be interpreted as the expected number of transitions from state i. Using these interpretations, we can give the formulas for the re-estimation of the parameters of the HMM as follows:

$$\begin{aligned}
\overline{\pi_i} &= \text{the probability of being in state } i \text{ at time } (t=1) \\
&= \gamma_t(i); \\
\overline{a_{ij}} &= \frac{\text{expected number of transitions from state } i \text{ to state } j}{\text{expected number of transitions from state } i} \\
&= \frac{\sum_{t=1}^{T-1} \zeta_t(i,j)}{\sum_{t=1}^{T-1} \gamma_t(i)}; \\
\overline{b_j(k)} &= \frac{\text{expected number of times in state } j \text{ and observing symbol } v_k}{\text{expected number of times in state } j} \\
&= \frac{\sum_{t=1\ O_t=v_k}^{T} \gamma_t(j)}{\sum_{t=1}^{T} \gamma_t(j)},
\end{aligned} \tag{3.25}$$

where $\overline{\pi_i}$, $\overline{a_{ij}}$ and $\overline{b_j(k)}$ are the re-estimated parameters of the model, and $\gamma_t(i)$ at $t = T$ can be calculated from the formula:

$$\gamma_t(i) = \frac{\alpha_t(i)\beta_t(i)}{\sum_{j=1}^{N} \alpha_t(j)\beta_t(j)}. \tag{3.26}$$

The above re-estimation procedure warrants that the given model can be improved iteratively, because it is always true that $Pr(\mathbf{O}|\overline{\lambda_w}) \geq Pr(\mathbf{O}|\lambda_w)$, where $\overline{\lambda_w}$ denotes the current model and λ_w denotes the model after re-estimation [9]. It should be noted that computing $\gamma_t(i)$ from (3.24) involves only additions if $t < T$, whereas directly computing $\gamma_t(i)$ from forward and backward variables costs more as it involves additions, multiplications and divisions.

B: The Viterbi Re-estimation Algorithm

Although the Baum-Welch re-estimation algorithm is the most popular algorithm for training HMMs, it can be replaced by a simpler and more effective algorithm based on the Viterbi decoding approach [199]. In the Viterbi re-estimation algorithm, the actual counts of events are accumulated at each state based on the Viterbi optimal path tracking instead of computing the expectations of events in the Baum-Welch algorithm. In order to explain the Viterbi re-estimation algorithm clearly, we define several actual accounts as follows:

$$\begin{aligned}
n(.) &= \text{number of training examples,} \\
n(ij) &= \text{number of transitions from state } i \text{ to state } j, \\
n(i.) &= \text{number of transitions from state } i, \\
n(.j) &= \text{number of transitions entering state } j, \\
n(.j)|_{t=1} &= \text{number of times in state } j \text{ at } t = 1, \\
n(O(j) = v_k) &= \text{number of events that the observed symbol is } v_k \\
&\quad \text{and the state } j \text{ occurs jointly.}
\end{aligned} \tag{3.27}$$

The re-estimation formulas for the Viterbi algorithm are:

$$\begin{aligned}
\overline{\pi_i} &= \frac{n(.j)|_{t=1}}{n(.)}, \\
\overline{p_{ij}} &= \frac{n(ij)}{n(i.)}, \\
\overline{b_j(k)} &= \frac{n(O(j)=v_k)}{n(.j)}.
\end{aligned} \tag{3.28}$$

The Viterbi re-estimation algorithm has been proved to converge to a proper characterization of underlying observation [155, 89] and can give comparable recognition performance to that trained by Baum-Welch algorithm [199]. Due to its simplicity and effectiveness, the Viterbi re-estimation algorithm is used in a broader class of grammars.

3.1.4 Continuous Observation Hidden Markov Models

In many real applications, the observations are continuous and vector-valued. Although these observations can be quantized into discrete symbols and characterized by discrete probability densities as discussed before, some distortions are inevitably introduced during the vector quantization process. Therefore, a better method is to compute the observation probabilities in an HMM directly from the continuous feature set. It is advantageous to use HMMs with continuous observation densities to avoid accumulating distortions caused by the vector quantization process.

A: The Baum-Welch algorithm

In order to model the continuous features in a consistent and efficient way, we have to place some restrictions on the form of probability density function. Usually, the probability density function in each state of an HMM is represented by a multivariate mixture density function:

$$b_j(O) = \sum_{l=1}^{L} c_{jl} \mathcal{N}(O, \mu_{jl}, \Sigma_{jl}), \tag{3.29}$$

where L is the number of mixtures in each state, c_{jl} is the mixture coefficient for the mth component in state j and $\mathcal{N}$ is a symmetric density function, typically a Gaussian mixture density function that has mean μ_{jl} and covariance matrix Σ_{ij} for the mth component in state j. In order for $b_j(O)$ to be a normalized probability density function, the mixture coefficients must be non negative and satisfy the stochastic constraint,

$$\sum_{l=1}^{L} c_{jl} = 1, \qquad 1 \leq j \leq N. \tag{3.30}$$

(3.29) can be used to arbitrarily accurately approximate any continuous probability density function if L is large enough. However, an excessive large number of L may result in the overfitting problem. A suitable L is problem dependent.

The re-estimation formulas for the three quantities c_{jl}, μ_{jl} and Σ_{jl} for mixture l in state$j$ have been derived by Liporace and Juang [161, 131]. In their method, the parameter re-estimation for continuous HMM is similar to that for the discrete HMM, except that the emission density $b_j(O)$ in its continuous form is used to replace $b_j(k)$ and the parameters for the lth mixture in state j must be re-estimated. Therefore, the re-estimation formulas for p_{ij} and π_i are identical to those used for the discrete HMM. In order to represent the re-estimation formulas concisely, we define an intermediate quantity as follows:

$$\xi_t(j,l) = \frac{\alpha_t(j)\beta_t(j)}{\sum_{i=0}^{N} \alpha_t(i)\beta_t(i)} \times \frac{c_{jl}\mathcal{N}(O_t, \mu_{jl}, \Sigma_{jl})}{\sum_{m=1}^{L} c_{jm}\mathcal{N}(O_t, \mu_{jm}, \Sigma_{jm})}. \tag{3.31}$$

This quantity is the probability of being in state j at time t with the lth mixture component accounting for the observation O_t. The sum of $\xi_t(j,l)$ over all L mixtures has the same form of $\gamma_t(j)$, therefore $\gamma_t(j)$ is a special case of $\xi_t(j,l)$ when $L = 1$. Now, the mixture coefficient, mean and covariance matrix can be re-estimated by

$$\begin{aligned}
\overline{c_{jl}} &= \frac{\sum_{t=1}^{T} \xi_t(j,l)}{\sum_{t=1}^{T}\sum_{m=1}^{L} \xi_t(j,m)}, \\
\overline{\mu_{jl}} &= \frac{\sum_{t=1}^{T} \xi_t(j,l)\cdot O_t}{\sum_{t=1}^{T} \xi_t(j,l)}, \\
\overline{\Sigma_{jl}} &= \frac{\sum_{t=1}^{T} \xi_t(j,l)\cdot (O_t-\mu_{jl})(O_t-\mu_{jl})^T}{\sum_{t=1}^{T} \xi_t(j,l)}.
\end{aligned} \tag{3.32}$$

Rabiner gave a heuristic interpretation of these formulas for the re-estimation of the mixture coefficient, mean and covariance matrix [209]. Briefly, the re-estimation formula for c_{ij} represents a ratio between the expected number of times the system is in state j using the lth mixture and the expected number of times the system is in state j. The formula for μ_{jl} is the weighted observation over the time when the system is in state j using the lth mixture. A similar interpretation is given for the re-estimation for covariance matrix Σ_{jl}.

B: Re-estimation using Viterbi algorithm

If we use the Viterbi algorithm, a simpler procedure can be used to re-estimate the emission parameters. Consider the case of a single mixture of Gaussian density. For a given observation sequence and model, each observation is assigned to a state that produces it on the optimal state path, which is obtained by examining the backtracking information. We denote O_t as the observation produced by the state i.

$$\begin{aligned}
\mu_{il} &= \frac{1}{N_{il}} \sum_{\substack{t=1 \\ O_t(i)}}^{T} O_t, \\
\Sigma_{il} &= \frac{1}{N_{il}} \sum_{\substack{t=1 \\ O_t(i)}}^{T} [O_t-\mu_{il}][O_t-\mu_{il}]^T,
\end{aligned} \tag{3.33}$$

where N_{il} is the number of observations assigned to the lth mixture in state i, and $l = 1$ in this case. When $L > 1$ mixture densities are used, the observations assigned to state i can be divided into L subsets using the

K-means algorithm with $K = L$ [208]. The mixture coefficient c_{il} can be re-estimated as

$$c_{il} = \frac{N_{il}}{N_i}, \tag{3.34}$$

where N_i is the number of observations assigned to state i.

3.2 Recognizing Handwritten Numerals Using Statistical and Structural Information

In this section, we apply the fundamental theory of HMM to modeling statistical information of handwritten numerals. The optimal state sequences, which are the by-products of statistical models, are used to model structural information of handwritten numerals to improve the performance of the system.

3.2.1 Statistical Modeling

Features

Closed contours are useful in representing 2-D patterns. Like skeletons, closed contours are efficient representations of 2-D patterns. Most importantly, features extracted from closed contours can be arranged as sequences. Furthermore, there are many methods that can be used to describe closed contours. More popular methods use coefficients of the Fourier descriptors [198] and the chain code features [139] to encode closed contours.

In our system, we use chain code-based features including locations, orientations and curvatures. The locations of short lines are the x-axis and y-axis coordinates. As shown in Figure 3.6(b), the orientation of segment k is defined as the direction from segment $k - 1$ to $k + 1$ and is encoded into one of the 16 directions shown in Figure 3.6(a). Although conventional chain codes [86] are easier to obtain, they are dependent only on two neighboring pixels, which makes them sensitive to noise [197]. In the following, we introduce a new method that is relatively insensitive to noise as the features are extracted from five-point segments with two overlapping points. The number of chain code-based features used in our method is smaller than that of the conventional chain code, which in turn reduces the computational cost. The curvature vector of segment i is defined as follows:

$$\{cx_i, cy_i\} = \{x_{oi} - x_i, y_{oi} - y_i\}, \tag{3.35}$$

where

$$x_{oi} = \frac{x_{i+1} + x_{i-1}}{2}, \qquad y_{oi} = \frac{y_{i+1} + y_{i-1}}{2}.$$

Figure 3.6(b) shows the parameters of a curvature and its orientation. The curvature vectors associated with a particular orientation are quantized into three codewords. Therefore, The curvature vectors and the orientations of the line segments form $16 \times 3 = 48$ codewords. In such a way, we can represent an input image by a set of vectors:

$$\mathbf{O} = \{O_1, O_2, \cdots, O_T\}, \tag{3.36}$$

where

$$O_i = \{(x_i, y_i), (D_i)\},$$

D_i is the index of the codebook and T is the length of observation sequence or feature vectors.

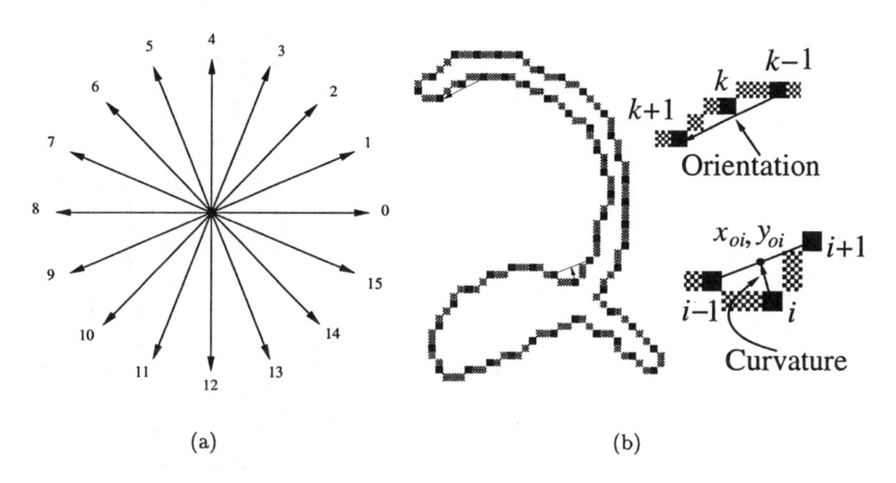

FIGURE 3.6. Feature representation. (a) Orientation codes. (b) Feature determination.

Left-Right Hidden Markov Models

The outer contours can be represented by sequential feature vectors, which are regarded as the equivalence of time-varying signals. The signals in short segments can be considered as stationary processes with minor fluctuations. Therefore, we can use individual states of the HMM to model the steady statistical information for short segments. Significant changes of properties over the whole input sequence can be modeled by state transitions. It is important to choose a right HMM because different types of HMMs are suitable for different types of signals. In our system, we use the left-right model, because in this way the features used in our system can be regarded as the equivalence of 1D time-varying signals. Further, the left-right model has the desirable property that it can readily model signals whose properties change over time [209]. An HMM with the left-right structure can be obtained by setting certain state transition coefficients to zero:

$$p_{ij} = 0, \qquad j < i \text{ or } j > i + \Delta \tag{3.37}$$

where Δ is a constant. In other words, no transitions are allowed for states whose indices are lower than the current state and large changes in state indices are prohibited. The main advantage of such HMMs is their ability to cope with variations in writing styles. Taking two '0's in Figure 3.7(a) for example, the two '0's can be well modeled by left-right HMMs with state sequences $\cdots j \rightarrow k \rightarrow q \rightarrow \cdots$ and $\cdots j \rightarrow o \rightarrow \cdots \rightarrow p \rightarrow q \rightarrow \cdots$, respectively. Two HMMs can be combined into one model shown in Figure 3.7(b). Therefore, the left-right type HMMs are able to deal with variations and uncertainties present in handwritten characters.

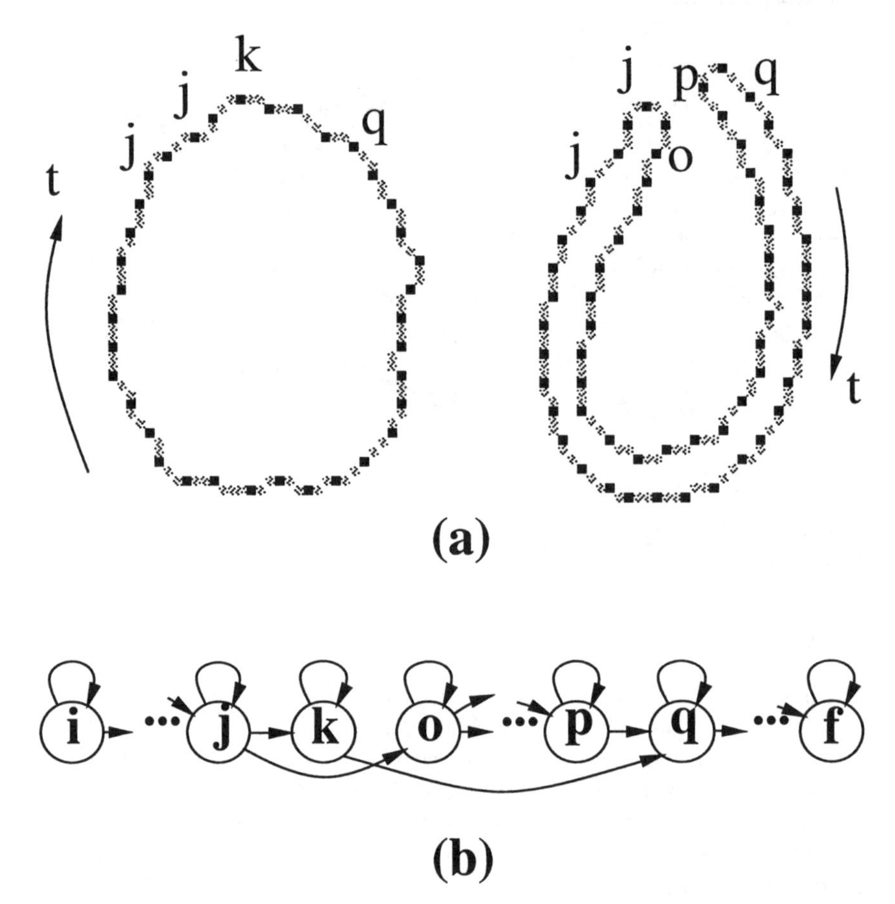

FIGURE 3.7. An example of using one left-right HMM to model a pattern class with huge variations.

Duration Modeling and Final State Modeling

The modeling of state duration is one major weakness of the conventional HMMs. The inherent duration probability density $p_i(d)$ associated with state i can expressed as follows [209]:

$$p_i(d) = a_{ii}^{d-1}(1 - a_{ii}). \tag{3.38}$$

With this model, the probability of the state duration decreases exponentially with its duration. Obviously, this exponential distribution of state duration is inappropriate for most physical signals. Several methods have been proposed to solve this problem [209, 157]; however, they require considerable computation. For example, in Levinson's method [157], $\alpha_t(j)$ is computed by

$$\alpha_t(j) = \sum_{i=0}^{N-1} \sum_{d=1}^{D} \alpha_{t-d}(i) a_{ij} p_i(d) \prod_{k=t-d+1}^{t} b_j(O_k), \tag{3.39}$$

where D is the maximum duration within any state. This method results in more than D times computation required in the standard HMMs. Vaseghi [247] proposed to use state distribution-depended transition probability to cope with this problem. The major advantage of Vaseghi's method is that it does not increase computation. As the model used in Vaseghi's method can skip over only one state and with our model we can skip over several states, we modify the state transition probabilities as:

$$a_{ii}(d_i) = 1 - \sum_{d=1}^{d_i} p_i(d), \tag{3.40}$$

$$a_{ij}(d_i) = \frac{N_{ij}}{\sum_{k>i} N_{ik}}[1 - a_{ii}(d_i)], \tag{3.41}$$

where N_{ij} is the number of state transitions from state i to state j. In addition, the numbers of feature vectors of the particular class may differ greatly from that of other classes, such as '1' and '8'. Therefore the word duration probability distribution can provide useful information for handwriting recognition.

In addition to duration probability distribution, the final state probability distribution that is not used in the conventional HMMs, can also be used to improve the performance of handwritten character recognition. The final state probability f_i is defined as the probability of the model ending at state i at time T, where $i \in S_F$.

Parameter Re-Estimation

We present a new approach to parameter re-estimation for HMMs for recognizing handwritten numerals. The Viterbi algorithm is adopted for training algorithm as it has the following important properties:

1. The state sequence of HMM is not observable. However, the single best state sequence can be found by the Viterbi algorithm. This is particularly useful in estimating state duration probabilities.

2. Using the Viterbi algorithm makes it possible to replace the multiplication in probability computation with summations. As a consequence, the scaling procedure, which is necessary in Baum-Welch algorithm, can be avoided. This results in better computational performance.

The standard Viterbi algorithm introduced in Section 3.1 is able to deal with state durations and final state probability distributions. In addition, multiplications in the Viterbi algorithm can be replaced by additions using logarithmic operations. Combining all parameters of an HMM and taking computation into consideration, we modify the Viterbi algorithm as follows:

Step 1: Initialization.

$$\alpha_1(i) = \log \pi_i + \log b_i(O_1),$$
$$\Phi_1(i) = 0 \qquad 0 \le i \le N-1,$$
$$d_i^1 = 1.$$

Step 2: Recursion. From time t=2 to T.

$$\alpha_t(j) = \max_i[\log \alpha_{t-1}(i) + \log a_{ij}(d_i^{t-1}) + \log b_j(O_t)],$$
$$\Phi_t(j) = \arg\max_i[\log \alpha_{t-1}(i) + \log a_{ij}(d_i^{t-1})],$$
$$d_j^t = d_i^{t-1} + 1 \qquad i = j, \text{ or } d_j^t = 1 \qquad i < j.$$

Step 3: Termination. (S_F is the final state set.)

$$\beta_1(w) = \log p^*(\mathbf{O}|\lambda_w) = \max_{s \in S_F}[\log \alpha_T(s) + \log f_s + \log p_w(d_w)],$$
$$q_T = \arg\max_{s \in S_F}[\log \alpha_T(s) + \log f_s + \log p_w(d_w)].$$

Step 4: State path backtracking. Form time t=T-1 to 1.

$$q_t = \Phi_{t+1}(q_{t+1}).$$

The re-estimation formulas of f_i, $p_i(k)$ and $p_w(k)$ are given as follows:

$$f_i = \frac{n(s_F = i)}{N_T}, \tag{3.42}$$

$$p_i(k) = \frac{n_s(d_i = k|s = i)}{\sum_h n_s(d_i = h|s = i)}, \tag{3.43}$$

$$p_w(k) = \frac{n_w(d_w = k)}{N_T} \tag{3.44}$$

where N_T denotes the number of training images belonging to the same pattern class, $n_s(d_i = k|s = i)$ the number of state i whose duration is k, $n_w(k)$ is the number of training images whose feature length $d_w = k$, and $p_w(k)$ the probability that the duration of the image belongs to class w is k. The initial state distribution can be re-estimated in the same way as that in the standard Viterbi algorithm and the state transition probabilities can be re-estimated by using (3.40) and (3.41).

In order to simplify the re-estimation of $b_j(O_t)$, we assume that the two components of O_t are independent. Therefore, we have

$$b_j(O_t) = p_{lj}(x_t, y_t) \cdot p_{cj}(k), \tag{3.45}$$

where

$$p_{cj}(k) = \frac{n(D_t = k|s = j)}{\sum_v n(D_t = v|s = j)}, \tag{3.46}$$

$$p_{lj}(x_t, y_t) = \frac{1}{2\pi|\Gamma_{lj}|^{1/2}} \exp\Big[-\tfrac{1}{2}(x_t - \overline{x}_j, y_t - \overline{y}_j)\Gamma_{lj}^{-1}\binom{x_t - \overline{x}_j}{y_t - \overline{y}_j}\Big], \tag{3.47}$$

$n(D_t = k|s_t = j)$ is the number of directions that are quantized as the codeword k when the state is j, $\overline{x}_j = E[x_t|s_t = j]$, $\overline{y}_j = E[y_t|s_t = j]$ and $\Gamma_{lj} = E[\binom{x_t - \overline{x}_j}{y_t - \overline{y}_j}(x_t - \overline{x}_j, y_t - \overline{y}_j)|s_t = j]$.

Initial Estimation of HMMs

The re-estimation procedure usually leads the initial models to their local optima, because the optimization surface is usually complex with local optima. Therefore, how to obtain initial models, which can result in global optimum, is a key issue in HMM-based approaches. Surprisingly, this important issue has received very little attention in the literature. There are many ways for the initialization of HMMs. Manual segmentation of observation sequences into states is a tedious and time-consuming procedure, which is inappropriate in most practical applications. Rabiner and his colleagues proposed the segmental k-means segmentation procedure [208]. However, this procedure needs its own initial models.

In this section, we will present a method for segmenting an observation sequence into states based on curvature. Let $c_t = \sqrt{cx_t^2 + cy_t^2}$, where cx_t and cy_t are defined in (3.35). The larger is the c_t, the sharper the curve of the segment. Therefore, we select N curve segments with the largest c_t. These short curves divide the closed outer curve into N segments (states). In this way, we obtain the initial models. Specifically, in our system, the complete initial models are obtained in three stages as shown in Figure 3.8.

As the complexity of the optimization surface is closely related to the number of model parameters: the more parameters, the more complex the surface. We do not attempt to obtain the initial models in a greedy way. Instead, in the first stage, we use only the curvature vectors and directions as features to obtain simpler models. This results in simpler optimization surface on which the re-estimation process will lead the initial models to the global optimal models with a higher probability. We will consider the locations and state durations in the second and third stages, respectively. Therefore, the final models are approached by re-estimation in succession.

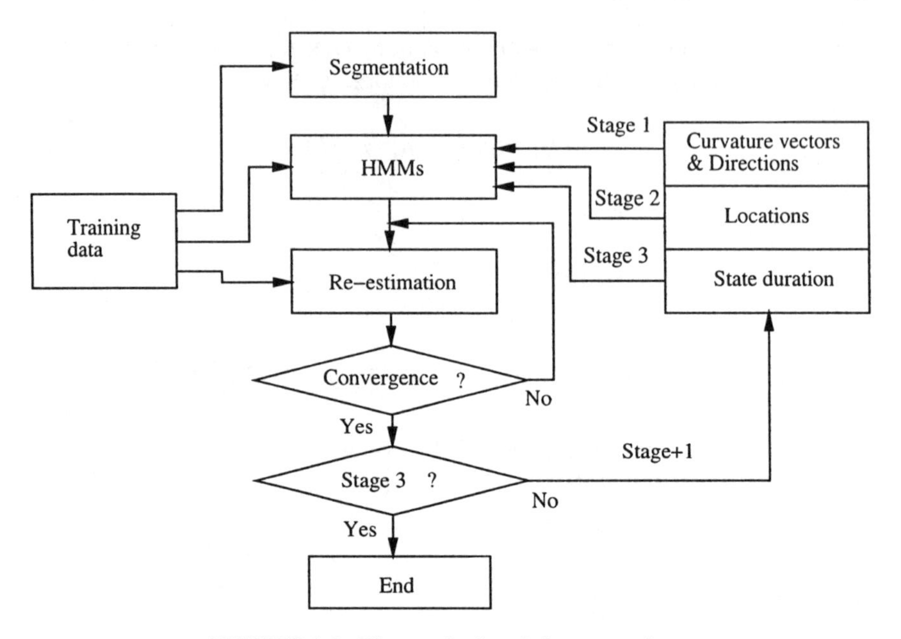

FIGURE 3.8. The gradual training procedure.

3.2.2 Structural Modeling

Another major weakness of HMMs is that it is difficult to model structural information in data. This is because the first-order Markov chain does not take the history of the state path into consideration. For a standard HMM, the next state is conditioned only by one preceding state:

$$p(q_t = j|\mathbf{X}) = p(q_t = j|q_{t-1} = i) = a_{ij}.$$

As a consequence, HMMs have difficulties to encode the history of the state path that can be used to model structural information of handwritten characters. In our system, we use a structural model to describe the relationship between states.

Structures can usually be modeled by relational graphs [215] or grammar representations [89]. One major drawback of the conventional structural approaches is that it is difficult to train prototype graphs or extract the grammars for structural representations [89]. In some structural approaches, the trained models have to be adjusted manually [183]. Due to shape variations in handwritten characters, it is impossible to use only one model to represent one class of handwritten characters in the conventional structural methods. Using multiple models may improve the recognition rate, but it slows down the recognition speed and increases the complexity of the system. These problems can be solved if we take the advantage of the HMM-based method's by-product: the single best-state sequence, from which we can establish *macro-states*.

Definition 3.2.1
A macro-state is defined as a collection of states with the same state index in the single best-state sequence.

A macro-state forms a state segment. The structure of a character can be modeled by parameters of these state segments (singletons) and the relationships between the state segments (two-nodes) involved. The similarity between the structure of the character and the reference model is measured by matching the structure of the character with the states in the best state sequence instead of the whole model.

Parameters of Macro States

Three parameters are used to describe a given macro-state i shown in Figure 3.9. The orientations are $d1_i$ and $d2_i$, where $d1_i$ is the direction from the starting point to the middle point on the curve, $d2_i$ from the middle point to the end point of the curve. However, the distributions of orientations cannot be directly modeled by Gaussian functions or Gaussian mixtures, because errors may occur in the estimation of average orientations due to discontinuities. Figures 3.10(a) and (b) show how errors can occur. The true orientation average of the two samples in Figure 3.10(a) is 0 or 360. However, if the range of orientations is [0,360], the estimated orientation average of the two samples is $(15+345)/2 = 180$, instead of near 0. Similarly, if the orientations are defined in the range of [-180, 180], the estimated orientation average of the two samples shown Figure 3.10(b) is $(-165+165)/2 = 0$, instead of 180. Therefore, the orientations are quantized into 16 codes shown in Figure 3.6(a). We define the location of the macro-state as the middle point on the curve denoted by a vector $\{X_{mi}, Y_{mi}\}$.

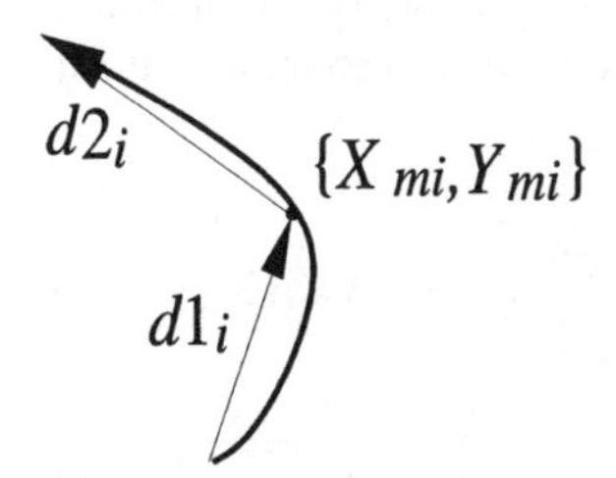

FIGURE 3.9. Singleton (single macro-state).

The relationships between macro-states are described by relative positions and relative orientations. For two macro-states (two-nodes) i and j $(i \neq j)$ shown in Figure 3.11(a), the relative position $\{X_{ij}, Y_{ij}\}$ is defined as the distance from the centroid of the macro-state i to that of the macro-state j:

$$\{X_{ij}, Y_{ij}\} = \{X_{mj} - X_{mi}, Y_{mj} - Y_{mi}\}. \tag{3.48}$$

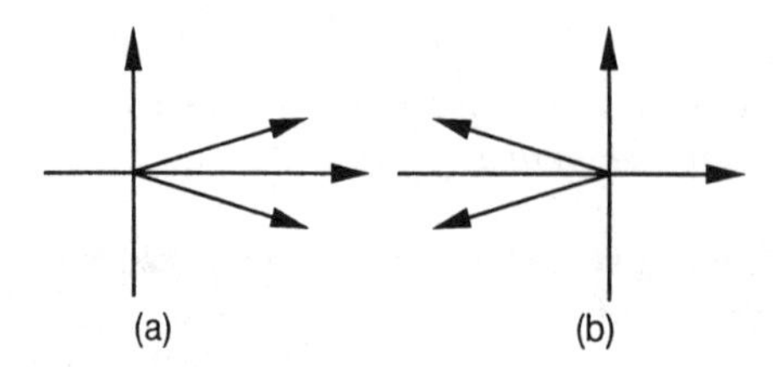

FIGURE 3.10. Examples of average orientation estimation errors.

The relative orientation d_{ij} between the two macro-states is defined as the angle between $d2_i$ and $d1_j$ measured anti-clockwise (See Figure 3.11(b)). The relative orientation is also encoded into one of the 16 orientation codewords in the codebook.

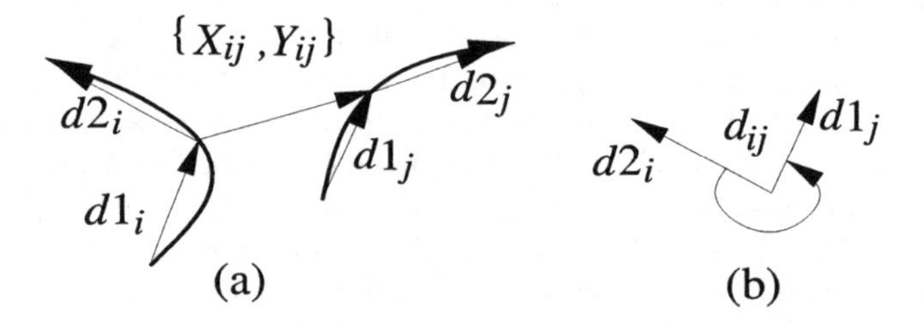

FIGURE 3.11. Relative orientation between two macro-states.

Statistical Structure Matching

The geometrical structures of handwritten words are described by parameters of macro-states and the relationships between macro-states. The recognition process is to measure the degree of matching or the cost of mismatching in terms of the structural description. The matching criterion is given by

$$\begin{aligned}\beta_2(w) = &\sum_{i=0}^{N-1} \mu_{1i} G_1(X_{mi}, Y_{mi}, d1_i, d2_i | \mathbf{M}_w) \\ &+ \sum_{i=0}^{N-1} \sum_{j=0, j\neq i}^{N-1} \mu_{2ij} G_2(X_{ij}, Y_{ij}, d_{ij} | \mathbf{M}_w),\end{aligned} \quad (3.49)$$

where μ_{1i} and μ_{2ij} are weights, $\mathbf{M}_w$ is the structural model of class w, $G_1(\cdot)$ is a matching function for a singleton, and $G_2(\cdot)$ is a matching function for a two-node pair. Now the problem of combining structural information becomes that of designing the matching function. Without loss of generality, we assign a larger value to the matching function when a structure matches well with the reference model. Obviously, such a matching function should be monotonically increasing. Gaussian and logarithmic Gaussian functions belong to such functions. Logarithmic Gaussian function is chosen in this chapter, as the logarithmic Gaussian function is able to model statistical information of structures and also consistent with the function used in the modified Viterbi algorithm described in Section 3.2.1. For simplicity, we assume that the orientations and the location of a macro-state are independent, and that the relative position and relative orientation are also independent. As a result, we may define the matching functions as follows:

$$G_1(X_{mi}, Y_{mi}, d1_i, d2_i|\mathbf{M}_w) = -\frac{1}{2}(X_{mi}-\overline{X}_{mi}, Y_{mi}-\overline{Y}_{mi})\chi_{1i}^{-1}\begin{pmatrix}X_{mi}-\overline{X}_{mi}\\Y_{mi}-\overline{Y}_{mi}\end{pmatrix}$$

$$-\frac{1}{2}\log|\chi_{1i}| + \log p_{1i}(d1_i) + \log p_{2i}(d2_i) - \log 2\pi, \qquad (3.50)$$

$$G_2(X_{ij}, Y_{ij}, d_{ij}|\mathbf{M}_w) = -\frac{1}{2}\log|\chi_{2ij}|$$

$$-\frac{1}{2}(X_{ij}-\overline{X}_{ij}, Y_{ij}-\overline{Y}_{ij})\chi_{2ij}^{-1}\begin{pmatrix}X_{ij}-\overline{X}_{ij}\\Y_{ij}-\overline{Y}_{ij}\end{pmatrix}$$

$$-\frac{1}{2}\log|\chi_{2ij}| + \log p_{ij}(d_{ij}) - \log 2\pi, \qquad (3.51)$$

where $\overline{X}_{mi}$, $\overline{Y}_{mi}$, $\overline{X}_{ij}$ and $\overline{Y}_{ij}$ are the means of X_{mi}, Y_{mi}, X_{ij} and Y_{ij}, respectively; χ_{1i} and χ_{2ij} are the covariance matrices and $p_{1i}(\cdot)$, $p_{2i}(\cdot)$, and $p_{ij}(\cdot)$ are discrete probability distribution functions. These parameters are given by

$$\chi_{1i} = E\left[\begin{pmatrix}X_{mi}-\overline{X}_{mi}\\Y_{mi}-\overline{Y}_{mi}\end{pmatrix}(X_{mi}-\overline{X}_{mi}, Y_{mi}-\overline{Y}_{mi})\right],$$

$$\chi_{2ij} = E\left[\begin{pmatrix}X_{ij}-\overline{X}_{ij}\\Y_{ij}-\overline{Y}_{ij}\end{pmatrix}(X_{ij}-\overline{X}_{ij}, Y_{ij}-\overline{Y}_{ij})\right],$$

$$p_{1i}(k) = \frac{n(d1_i = k)}{\sum_h n(d1_i = h)}, \qquad 0 \le k, h < N$$

$$p_{2i}(k) = \frac{n(d2_i = k)}{\sum_h n(d2_i = h)},$$

$$p_{ij}(k) = \frac{n(d_{ij} = k)}{\sum_h n(d_{ij} = h)},$$

where $n(d* = k)$ denotes the number of macro-state orientations that are encoded by codeword k.

3.3 Experimental Results

To evaluate the proposed approach, we used a standard subset of CEDAR CDROM1 database, which consists of 18,468 digit images for training and 2711 digit images for testing. In the test database (bindigis/bs), there is a subset (goodbs) consisting of 2213 well-segmented digit images. These data were collected from live mails at the main post office in Buffalo, NY. Figure 3.12 shows some representative samples from the database.

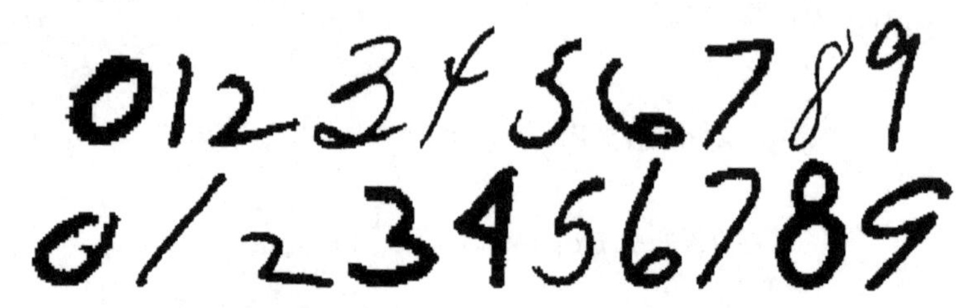

FIGURE 3.12. Some samples in the database.

In our experiments, the HMMs were trained by the modified Viterbi algorithm and the structural models were obtained from the best state sequences of HMMs that optimally decode the training data. The recognition of unconstrained handwritten numerals was performed by

$$w^* = \arg\max_{w}\{\beta_1(w) + \beta_2(w)\}. \tag{3.52}$$

During the training, we found that the selection of state numbers was crucial to structural models. If the state numbers are too large, the macro-state lengths may become short and the parameters for the macro-states may become unreliable. On the other hand, under-segmentation will reduce models' discriminant ability for different handwritten numerals of similar shapes. Therefore, the state number is different from model to model. If a character has a short outer contour, such as '1', it should be modeled by a small number of states; on the other hand if a character has longer outer contour, such as '8', it should be modeled by more states. In our experiments, we determine the state numbers by maximizing the recognition rate of the training set as shown in Table 3.1.

TABLE 3.1. The state numbers of models.

Model	0	1	2	3	4	5	6	7	8	9
State Number	14	8	14	14	12	14	12	12	16	12

We conducted our experiments on the "bindigis" set and the "goodbs" set. The performance of the recognizer with and without structural models (SMs) is compared in Table 3.2. If the recognizer is based on HMMs only, the correct recognition rates on these two sets were 95.46% and 98.01% respectively. However, the structural models can reduce the recognition errors by 15.42% and 18.09% respectively. Because the "goodbs" set is well segmented, the performance on this test set can exactly reflect the characteristics of the proposed approach. Table 3.3 gives the confusion matrix of the recognizer based on statistical-structural models (SSMs) for "goodbs" set. From the confusion matrix, we find that there are several confusion pairs of handwritten numerals such as "4-9", "0-8" and "6-4". As the features were extracted from outer contours of images, the shapes of these pairs are similar; and it is not surprising that these pairs are relatively difficult to distinguish. Nevertheless, the performance of this approach is impressive. Our experimental results achieved by one model per class are comparable to the best results published recently [154, 59, 183, 120, 103]. Some other approaches use several models per class or combine several classifiers to recognize handwritten numerals. For instance, Hwang and Bang use 50 clusters per class to achieve 97.90% accuracy [120]. Ha and Bunke [103] and Lee [154] combine several classifiers with different input features to achieve 99.09% and 97.1% recognition rates.

The average recognition speed measured on an old Silicon Graphics workstation ($4 \times\ 33MHz$ IP7) is 7 digits per second.

TABLE 3.2. Comparison of performance with and without SMs.

Test Set	Recognition Rates		Error Rates		Error Reduction
	HMMs	SSMs	HMMs	SSMs	(SSMs vs HMMs)
bindigis	95.46%	96.16%	4.54%	3.84%	15.42%
goodbs	98.01%	98.37%	1.99%	1.63%	18.09%

TABLE 3.3. Confusion matrix for "goodbs" set; the last column shows the recognition rates in percentage.

Total	Class	0	1	2	3	4	5	6	7	8	9	Rate %
355	0	349	3							3		98.31
289	1		289									100.00
224	2			220	3					1		98.21
208	3			2	205		1					98.56
183	4					178		1			4	97.27
117	5				2		114			1		97.44
245	6	1				3		241				98.37
221	7			1	1				217		2	98.19
191	8	3	1		1					185	1	96.86
180	9								1		179	99.44
2213	ALL											98.37

3.4 Conclusion

In this chapter, we have introduced HMMs and presented a novel approach to unconstrained handwritten numeral recognition, which is able to integrate statistical and structural information. The success of this system lies in that the features extracted from perfect outer contours can be arranged into sequential vectors that are suitable for HMMs; and segmentation by macro-states avoids inconsistent segmentation problems in traditional structural approaches and the time-consuming multi-to-multi matching. The structural similarity is measured by matching the structure of a test image to the structure of the best macro-state sequence in the model; therefore one model is able to cope with large shape variations in handwritten numerals belonging to the same class. The experimental results show that the proposed approach based on statistical-structural models can achieve high performance in terms of recognition speed and accuracy. It is expected that performance of the approach described in this chapter can be further improved if some global features are used and multiple models per class are adopted.

4

Markov Models with Spectral Features for Handwritten Numeral Recognition

Probabilistic models are powerful in coping with large variations in shapes, whereas Fourier spectra are suitable for describing 2-D shapes with simple closed curves. In this chapter, we introduce a hybrid recognition method that uses Markov process to model spectral features for recognizing handwritten numerals. We analyze the properties of the Fourier descriptors for spectral features derived from contours of 2-D shapes. These features can be used for 2-D pattern recognition. Section 4.1 gives a brief review of 2-D pattern recognition using the features extracted from contours. Section 4.2 introduces Fourier descriptors for spectral features. Section 4.3 presents the Markov model-based method for recognizing handwritten numerals. This chapter also presents efficient re-estimation and evaluation algorithms. The results of handwritten numeral recognition using the proposed method are given in Section 4.4.

4.1 Related Work Using Contour Information

Contour-based features have been used in many pattern recognition systems, because they have several advantages. Representing patterns by contours can reduce data storage comparing with original images. Another important advantage is that contour information is suitable to represent 2-D shapes by various techniques. Pavlidis and Ali proposed to approximate shapes of handwritten numerals by polygons [197]. In their system, the features were generated by polygonal approximations, the recognition decision was made by using classification trees. They were able to achieve a accuracy of 90.6% for handwritten numeral recognition. However, it is difficult to automatically obtain the classification trees and the recognition based on semantics was sensitive to shape variations and noise. Fourier descriptors have been widely used in pattern recognition as the absolute values of the descriptors are invariant to rotation and reflection [178, 135], which is a desirable property in some applications such as hand-tool recognition and shape-based image retrieval [223, 224]. However, such systems have difficulties in recognizing some pairs of characters such as '6' vs. '9' and '2' vs. '5', as these are rotated versions of each other. The auto-regressive (AR) models have been extensively used in speech analysis and speech recognition [69, 172]. Recently Dubios *et al.* [74], Sekita *et al.* [225] and others [53, 65] used the auto-regressive models to describe contours of 2-D patterns. They recognize patterns by using weighted Euclidean distances between auto-correlation coefficients of prototypes and test patterns, where 1-D signals were obtained by sampling points on contours in two ways as shown in Figure 4.1. As the auto-correlation coefficients, which are used to derive AR models, are insensitive to phase, the AR descriptors can be invariant to rotation and starting points of contours. However, AR contour descriptors suffer from the same problems as that in [223, 224] for handwriting recognition.

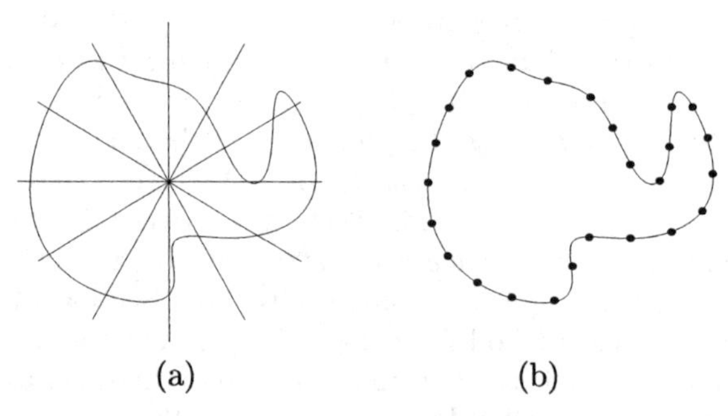

FIGURE 4.1. Contour point sampling. (a) Polar coordinate expression. (b) Complex coordinate expression.

In an attempt to solve these problems, Persoon *et al.* and Cheng *et al.* proposed to use complex Fourier descriptors for recognizing handwritten numerals [198, 51]. In complex Fourier descriptors, the Fourier coefficients are sensitive to the choice of starting points; different starting points result in different spectra of the same shape. Persoon and Fu obtained the optimal similarity matching by adjusting starting points in a certain range [198]. A major disadvantage of this method is that the similarity measures cannot deal well with large variations in shapes. As a result, Persoon and Fu obtained a recognition rate of only 89.4% for handwritten numerals. Cheng and Yan dealt with the phase by selecting starting points according to a heuristic rule [51]. In their system, a tree classifier was used to recognize handwritten numerals based on the summation of Euclidean distances of complex Fourier coefficients between prototypes and test patterns. However, there are three major problems in their method: First, the slants of handwritten digits change the values of the complex Fourier coefficients, which may result in significant mismatch between the prototypes and test patterns; second, their method is unable to deal with large variations of shapes using few prototypes or models per digit. Therefore, they had to use more than 250 prototypes in their system in order to obtain good results. Furthermore, selecting the representative prototypes from the training data set is a difficult task. Portegys suggested to use all the samples in the training set as prototypes [203]. However, this method is not feasible for problems needing a large training database. An immediate method would be to obtain the representative prototypes manually from each class. Unfortunately, there are some difficulties with this method: 1) it is tedious and time consuming to select representative prototypes manually; 2) there is no guarantee that this procedure will result in a set of optimized prototypes per class; 3) it requires intensive computation to evaluate each unknown sample with respect to the prototypes.

In real-world applications, it is desirable to use one model per class for representing all possible deformations in the training set. The HMM is a good choice to achieve this goal as it can deal with large variations; however, it requires sequential features as its inputs. Fourier descriptors based on contours are suitable for representing 2-D shapes [135] and can convert 2-D shapes into sequential features in spectral space. Therefore, using HMMs to model features generated by Fourier descriptors is advantageous for recognizing handwriting. In this chapter, we will present a new statistical approach that is able to use one model per class to efficiently recognize unconstrained handwritten numerals.

4.2 Fourier Descriptors

Contour information is useful in both data compression and pattern representation for 2-D shapes. Many contour descriptors have been developed for reconstructing and classifying 2-D shapes, such as complex autoregressive (CAR) descriptor [225], complex Fourier descriptor [178], complex PARCOR descriptor [225], affine-invariant Fourier descriptor [3] and moment-invariant descriptor [16, 206], etc. Sekita *et al.* compared the performance of several different descriptors based on contours [225]. Their experimental results show that complex AR descriptors can indeed achieve better recognition rates than other descriptors with low feature dimensions, and that the complex Fourier descriptor can obtain high recognition rates with high feature dimensions. Kauppinen *et al.* compared the performance of the autoregressive and Fourier descriptors in 2-D shape classification [135]. They carried out experiments on recognizing airplanes and machine-printed letters with rotation and affine transformations and noise. They concluded that complex Fourier descriptor and affine-invariant descriptor based on contours are the most effective descriptors, especially when perspective distortion and noise are present.

It is well known that Markov models are able to cope with large variations [37], while the complex Fourier descriptor based on contours is able to achieve high performance for rigid 2-D shape recognition. Therefore, the combination of the Fourier descriptor and Markov models is highly desirable in recognizing non-rigid 2-D shapes such as handwritten numerals.

Let us assume that C is a simple, clockwise closed contour and there are N points evenly sampled over the complete contour cycle. These points are characterized by their complex coordinates:

$$Z = \{z_i\}, \qquad 0 \leq i \leq N-1 \tag{4.1}$$

where

$$z_i = (x_i - x_c) + j(y_i - y_c), \tag{4.2}$$

in which x_i and y_i are the coordinates of ith sampled point on the contour, $j = \sqrt{-1}$, x_c and y_c are the coordinates of the centroid of the contour. In our experiments, N is set to an exponential function of 2 so that FFT can be used to reduce the computation of the Fourier spectrum. The complex spectrum of a shape can be represented as

$$F = \{F_0, F_1, \cdots, F_i, \cdots, F_{N-1}\}, \tag{4.3}$$

where F_i is the component of the spectrum at frequency i and can be calculated by

$$F_i = \sum_{k=0}^{N-1} z_k \exp(-j2\pi ik/N). \tag{4.4}$$

For the conventional Fourier descriptors, the representative prototype is selected from data belonging to the same class and the distance is based on direct matching between the prototype and the testing pattern. Persoon and Fu defined the optimal distance as follows [198]:

$$d_{FO} = \min_s \min_\alpha \min_\phi \sum_{i=-N/2+1, i\neq 0}^{N/2} |F_i - sO_i e^{j(i\alpha+\phi)}|^2, \tag{4.5}$$

where $F_{-i} = F_{N-i}$, O is the spectrum of the testing pattern, s is size scale, α is a factor for the starting point, and ϕ is caused by rotation. These Fourier descriptors with simple Euclidean distances have two main disadvantages: First, one model or prototype based on the distance in (4.5) cannot represent more than one shape for the same class. Figure 4.2 illustrates such examples. The intra-class distances of Fourier coefficients between '0_1' and '0_2' and between '4_1' and '4_2' are 1.652 and 1.530, respectively. However, the inter-class distance between '0_2' and '4_1' is 1.356 that is smaller than the intra-class distances. A simple solution to this problem is to use multiple prototypes or models [51], which results in increased computation.

The second main disadvantage is that the definition of the optimal distance in (4.5) cannot be used for maximizing the likelihood within the samples of similar shapes. If there are some training samples of handwritten numerals having similar shapes, we can obtain a prototype by taking the average of the spectra over the training samples after phase and scale

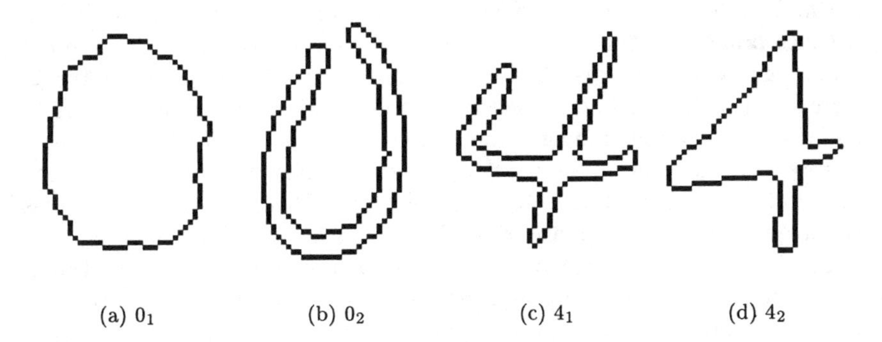

(a) 0_1 (b) 0_2 (c) 4_1 (d) 4_2

FIGURE 4.2. The variations of handwritten characters in shape.

normalization. However, the variance of each spectral component, which can be used to maximize the likelihood within the training samples, is not taken into consideration.

Nevertheless, the Fourier descriptors are able to convert the 2-D representation of a pattern into a 1-D problem. This allows us to employ many well-established techniques for 1-D signal processing, such as HMMs, for 2-D shape recognition. Furthermore, the Fourier spectrum has very clear physical meanings, namely, the lower frequency part of the spectrum represents the information of the general shape, and the higher frequency part of the spectrum contains the information about the details of the shape. In addition, the phase of Fourier spectrum represents the information about rotation, reflection and the starting point. In the following sections, we present a hybrid method that can fully utilize the advantages of Markov models and the Fourier descriptors for recognizing unconstrained handwritten numerals.

4.2.1 Feature Extraction

In order to obtain a consistent set of features and achieve high recognition rates for handwritten numeral recognition, we need to pre-process images to reduce variations. Some of the major pre-processing techniques relevant to this book have been described in Chapter 2, which include the connection of disjoint regions and slant correction. After the pre-processing stage, features for handwriting recognition are extracted from the outer contours of handwritten images using the complex Fourier transform. Figure 4.2 shows some examples of the outer contours of slant-corrected images.

Using contour-based Fourier spectra as features has several advantages, the most noticeable of which is that the magnitude of the Fourier spectrum is robust to affine transformation in a small range (0°-30°). Experimental results have shown that the k-nearest neighbor (KNN) classifier using the contour-based Fourier spectrum can achieve the best or close to optimal performance if the range of affine transformation is small [135], because affine transformation (including slant transformation) in a small range can be approximated by rotation. This is particularly useful in recognizing handwritten numerals as the slant estimation method is not perfect and the slant-corrected images may have residual slant. Further, the effects of rotation (actually minor slant) and the starting point exist only in the phase of the contour-based Fourier spectrum. Therefore, we can process the magnitude and the phase of the Fourier spectrum independently.

The first step in feature extraction is to calculate the contour-based complex Fourier spectrum using (4.4). However, the spectrum is not invariant to scale. In order to achieve scale invariance, we set the length of the closed contour to a constant, say, L_c. Therefore, the normalized spectrum $\tilde{F}$ can be expressed as follows:

$$\tilde{F} = \{\tilde{F}_0, \cdots, \tilde{F}_i, \cdots, \tilde{F}_{N-1}\} = \{\frac{F_0 L_c}{L}, \cdots, \frac{F_i L_c}{L}, \cdots, \frac{F_{N-1} L_c}{L}\}, \quad (4.6)$$

where L is the length of the original contour. As the influence of the object position is removed from the complex coordinates of contour points in (4.2), the spectrum in (4.6) is invariant to translation. This is equivalent to removing $\tilde{F}_0$ from the spectrum.

Now, we consider the influence of the starting point, rotation and reflection. For convenience of future analysis, we express $\tilde{F}_i$ as $|\tilde{F}_i|e^{j\theta_i}$, where $|\tilde{F}_i|$ is the magnitude part of the spectrum at frequency i and θ_i is its phase. Let $\hat{F}$ denote the spectrum after adjusting the starting point by k points and rotation by an angle of ϕ, we have

$$\hat{F}_i = |\tilde{F}_i|e^{j(i\alpha+\phi+\theta_i)}, \quad (4.7)$$

where the phase at frequency i is $\Theta_i = i\alpha + \phi + \theta_i$, $\alpha = 2\pi k/N$ and k is related to the starting point. From (4.7), we can find that the starting point and the rotation change only the phase of the spectrum. This means that the magnitude $|\tilde{F}|$ of the spectrum is invariant to the starting point, scale, translation and rotation, and the phase spectrum contains the information of the starting point and rotation. As different starting points result in different phase spectra of the contour, we must select starting points based on a certain criterion. In this system, we first choose (x_c, y_{top}) as a reference point, where x_c is the x-coordinate of the image centroid and y_{top} is the y-coordinate of the top point of the image. We choose the starting point as the pixel on the outer contour having the longest distance from the reference point. From (4.7), we find that the effect of the starting point $i\alpha$ is a linear function of the frequency i. Usually, $k \ll N$ and the effect of $i\alpha$ on the phase spectrum is not significant when the frequency is very low. However, if the frequency is high (close to $i = N/2$), $i\alpha$ makes the phase spectrum meaningless. Therefore directly using phases as features is not appropriate. Instead, we are interested in the delta phase spectrum:

$$\Delta\Theta_i = \Theta_i - \Theta_{i-1} + \delta_i = \theta_i - \theta_{i-1} + \alpha + \delta_i, \quad (4.8)$$

where $\Theta_0 = 0$, δ_i is the factor to ensure the continuity of the phase spectrum and is determined by

$$\delta_i = \begin{cases} 0 & \text{if } |\Theta_i - \Theta_{i-1}| \leq \pi, \\ 2\pi & \text{if } \Theta_i - \Theta_{i-1} < -\pi, \\ -2\pi & \text{if } \Theta_i - \Theta_{i-1} > \pi. \end{cases} \tag{4.9}$$

The delta phase spectrum together with the magnitude spectrum forms a set of complete features $\mathbf{O}$ for recognizing handwritten numerals

$$\mathbf{O} = \{(M_1, \Delta\Theta_1), (M_2, \Delta\Theta_2), \cdots, (M_{N-1}, \Delta\Theta_{N-1})\}, \tag{4.10}$$

where the magnitude of the contour-based Fourier spectrum is denoted as

$$M = \{M_i\}, \qquad 1 \leq i \leq N-1 \tag{4.11}$$

and $M_i = |\tilde{F}_i|$.

4.3 Hidden Markov Model in Spectral Space

4.3.1 Spectral Space

As there are two elements in each frequency component, the straightforward way of representing the output probability of each frequency component is to use a joint probability density function (*pdf*), such as joint Gaussian function. However, the joint *pdf* cannot be used to model the relationship between M_i^t and $\Delta\Theta_i^t$. To deal with this problem, we use another set of state transition probabilities to describe the relationship between M_i^t and $\Delta\Theta_i^t$. To formally describe HMMs in the spectral space, we define the following notations:

- K is the number of states for one component of the Fourier magnitude spectrum.
- $\mathbf{X}$ is a set of states $\{s_{i,j}\}$, where $s_{i,j}$ denotes the jth state for the ith component of the magnitude spectrum.
- $\mathbf{A}$ is a state transition probability distribution, where

$$\mathbf{A} = \{a_{i,j,k}\}$$

  where $a_{i,j,k} = Pr(s_{i+1,k}|s_{i,j})$ and denotes the state transition probability from state $s_{i,j}$ to state $s_{i+1,k}$.
- $\mathbf{B}$ are the output probability density functions for given states and observations, where

$$\mathbf{B} = \{b_{i,j}(M_i) = Pr(M_i|s_{i,j})\}.$$

- π is the initial state distribution, where
$$\pi = \{\pi_j\},$$
in which $\pi_j = Pr(s_{1,j})$.

- $\hat{\mathbf{X}}$ is a set of states $\{\hat{s}_{i,j}\}$ for modeling the delta phase spectrum, where $\hat{s}_{i,j}$ denotes the jth state for ith component of the delta phase spectrum, $1 \leq i \leq N-1$ and $1 \leq j \leq K$.

- $\hat{\mathbf{A}} = \{\hat{a}_{i,j,k}\}$ is a state transition matrix, where we assume that $\Delta\Theta_i$ is dependent on M_i for simplicity, and $\hat{a}_{i,j,k}$ denotes the state transition probability from state$s_{i,j}$ to state $\hat{s}_{i,k}$.

- $\hat{\mathbf{B}}$ are the output probability functions for the delta phase spectrum, where
$$\hat{\mathbf{B}} = \{\hat{b}_{i,k}(\Delta\Theta_i) = Pr(\Delta\Theta_i|\hat{s}_{i,k})\}.$$

Correspondingly, we can use the compact notation, $\lambda = (\mathbf{A}, \hat{\mathbf{A}}, \mathbf{B}, \hat{\mathbf{B}}, \pi)$, to represent the proposed model; and its structure is shown in Figure 4.3.

Note that there are significant differences between the HMMs with spectral features and the conventional HMMs in addition to the differences in the definitions and the model structures. In modeling signals in the time domain, we can use one state to model the signal that spans a number of time frames (segments) in which it remains statistically stationary. Due to the duality nature of the signal and its spectrum, it is meaningless to use one state of an HMM to model the statistics of continuous spectral components: if a state were used to model several continuous spectral components, a small difference of state duration would result in large variations in shapes. Therefore the state duration of the proposed HMM should always be one (one component). State transitions are used to model the changes of statistics in signals in the time domain. Meanwhile, the state transitions of HMMs in the spectral space are used to model the constraints between neighboring components or the relationships between magnitude and phase components of the Fourier spectra.

For Markov model-based methods, the recognition decision is based on the output probability $Pr(M, \Delta\Theta|\lambda)$ that is the summation of the joint probabilities over all possible state sets that M, $\Delta\Theta$, $\mathbf{X}$ and $\hat{\mathbf{X}}$ occur simultaneously:

$$Pr(M, \Delta\Theta|\lambda) = \sum_{\text{all } \mathbf{X}, \hat{\mathbf{X}}} Pr(M, \Delta\Theta|\mathbf{X}, \hat{\mathbf{X}}, \lambda) Pr(\mathbf{X}, \hat{\mathbf{X}}|\lambda), \tag{4.12}$$

where $Pr(M, \Delta\Theta|\mathbf{X}, \hat{\mathbf{X}}, \lambda)$ is the probability of observations M and $\Delta\Theta$, given state sets $\mathbf{X}$, $\hat{\mathbf{X}}$ and model λ. For computational feasibility, we assume that the Markov model for modeling the magnitude of the Fourier spectrum

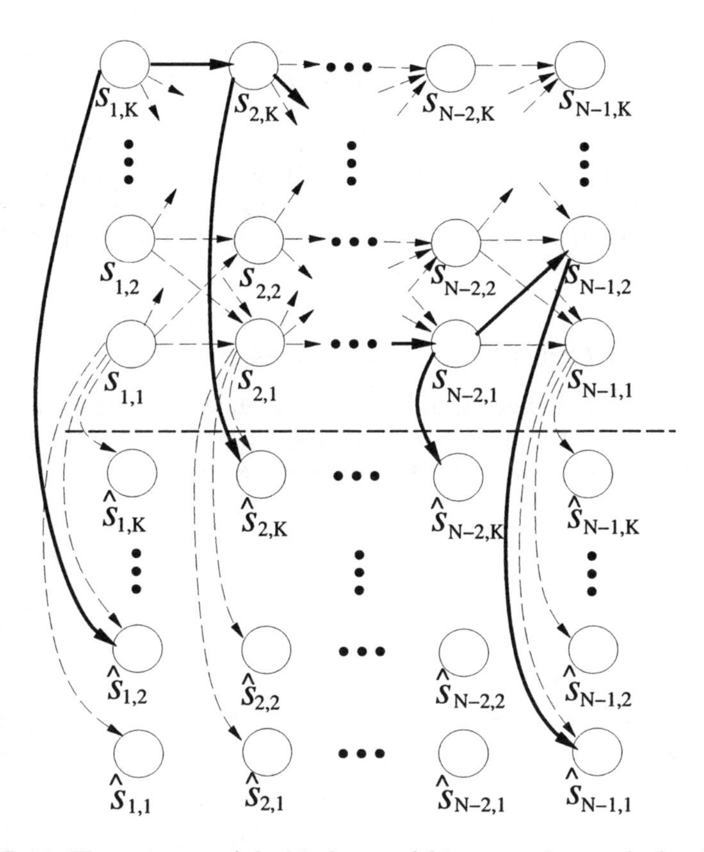

FIGURE 4.3. The structure of the Markov model in spectral space for handwritten numeral recognition.

is of first order, the states for modeling the phase spectrum is dependent only on the states for modeling the magnitude spectrum with the same frequencies, and that the probabilities of the observations for given states are independent. As the model structure is causal, the probability of an observation sequence for a given state sequence can be expressed as follows:

$$Pr(M, \Delta\Theta|\mathbf{X}, \hat{\mathbf{X}}, \lambda) = \pi_{q_1} Pr(M_1|s_{1,q_1})\hat{a}_{1,q_1,\hat{q}_1} \tag{4.13}$$

$$Pr(\Delta\Theta_1|\hat{s}_{1,\hat{q}_1} \cdots a_{N-2,q_{N-2},q_{N-1}}$$

$$Pr(M_{N-1}|s_{N-1,q_{N-1}})\hat{a}_{N-1,q_{N-1},\hat{q}_{N-1}}$$

$$Pr(\Delta\Theta_{N-1}|\hat{s}_{N-1,\hat{q}_{N-1}}), \tag{4.14}$$

where q_i and $\hat{q}_i$ are state indices for the magnitude and phase parts, respectively. This causal property allows us to use one-pass algorithm to find the global maximum of the output probability for the given model and observation.

4.3.2 Semi-Continuous Markov Model

As we discussed previously, in handwritten numerals there are significant variations in the contour-based Fourier spectra of the same class due to large shape variations. Figure 4.4 shows such an example.

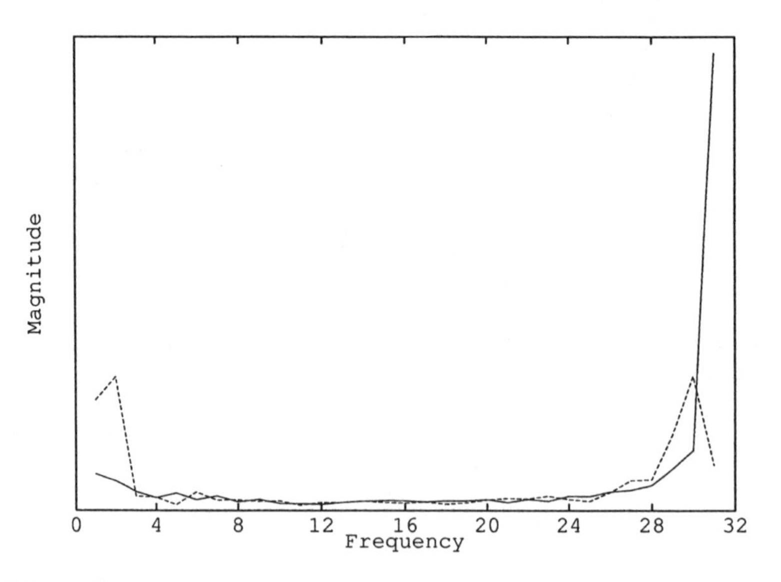

FIGURE 4.4. The variations in contour-based Fourier spectra of handwritten numerals, where the solid line is the magnitude of '0_1' and dash line is the magnitude of '0_2'.

If we use the continuous HMMs (CHMMs), in which the observations are modeled by continuous probability density functions [161, 131, 132], to deal with the variations in spectrum, we need a mixture of probability functions to make a good approximation of the feature distribution per state. However, this will increase computational complexity considerably as well as the number of model parameters. More importantly, such a model is very sensitive to initial conditions.

On the other hand, with vector quantization we can generate a codebook from training data and represent each observation by the closest codeword in the codebook in terms of the discrete HMMs (DHMMs) [100, 169]. One of the most important properties of DHMMs is that the discrete distributions are able to model observations with any distribution if there are enough training data. However, since it quantizes continuous features into discrete codewords, DHMMs may result in serious quantization errors.

In order to overcome these disadvantages, researchers proposed semi-continuous HMMs (SCHMMs) [112, 113, 111] that combines the advantages of CHMMs and DHMMs. In SCHMMs, we use a mixture of continuous probability density functions associated with the vector quantization (VQ) codebook to model the output probability of a state, where the VQ codebook can be obtained by the techniques proposed in [63, 212]. From the viewpoint of CHMMs, SCHMMs reduce the number of free model parameters, because continuous probability density functions are associated with the VQ codebook. From the viewpoint of DHMMs, the continuous probability density functions can significantly minimize the errors caused by the standard VQ [100, 169]. Experiments on speech recognition have shown that SCHMMs can achieve better results than DHMMs and CHMMs [111].

However, SCHMMs must be modified in order to deal with the problems in handwritten numeral recognition. In the standard SCHMMs, the VQ codebook represents the features from the whole training data. However, such a VQ codebook is not directly applicable to this case, because the features at different frequencies have different physical meanings. In our approach, we use one state to model the feature statistics at a specific frequency. Since the magnitude and phase spectra also require different VQ codebooks, in this chapter, we use two codebooks per class at a *specific* frequency for the magnitude and the phase components of the spectra. This results in $2N$ codebooks for each class, where each codebook is created specially for one type of features at a specific frequency per class. Let V denote the number of codewords or clusters in each codebook, C_i be the codebook for the magnitude spectrum at frequency i, and $\hat{C}_i$ be the codebook for the phase spectrum at frequency i. The probability density functions that produce M_i and $\Delta\Theta_i$ can be expressed as follows:

$$b_{i,j}(M_i) = Pr(M_i|s_{i,j}) = \sum_{v=1}^{V} f(M_i, c_{i,v}) Pr(c_{i,v}|s_{i,j}) \quad (4.15)$$

and

$$\hat{b}_{i,j}(\Delta\Theta_i) = Pr(\Delta\Theta_i|\hat{s}_{i,j}) = \sum_{v=1}^{V} f(\Delta\Theta_i, \hat{c}_{i,v}) Pr(\hat{c}_{i,v}|\hat{s}_{i,j}), \quad (4.16)$$

where $f(\cdot,\cdot)$s are functions associated with the codebooks, and $c_{i,v}$and $\hat{c}_{i,v}$ are vth codewords in C_i and $\hat{C}_i$, respectively.

4.3.3 Evaluation, Re-Estimation and Initiation

For a given model, there are three key problems which must be solved:

1. Given the observation sequence $\{M_i, \Delta\Theta_i\}$ and model λ, how do we compute the output probability?
2. Given the model λ and the observation sequence $\{M_i, \Delta\Theta_i\}$, how do we adjust the model parameters to maximize the output probability?
3. Given a training set, how do we obtain an initial model?

A: Evaluation

The first problem is an evaluation problem. The load for direct computation of (4.12) is still very heavy even for a first-order HMM. We modify the standard Viterbi algorithm for evaluation. Our HMM has a 2-D structure. For a given state $s_{i,k}$, we need to find not only the state $s_{i-1,j}$ from which the state transition is optimal, but also the state $\hat{s}_{i,m}$ to which the state transition is optimal. At the same time, we need to record the state information in terms of magnitude and phase spectra. Given these constraints, we modified the Viterbi algorithm as follows:

Step 1: Initialization.

$$\alpha_1(j) = \max_m [\hat{a}_{1,j,m}\hat{b}_{1,m}(\Delta\Theta_1)]\pi_j b_{1,j}(M_1),$$
$$\Phi_1(j) = 0 \qquad 1 \le j \le K,$$
$$\hat{\Phi}_1(j) = \arg\max_m [\hat{a}_{1,j,m}\hat{b}_{1,m}(\Delta\Theta_1)], \qquad 1 \le m \le K.$$

Step 2: Recursion.

$$\alpha_i(k) = \max_j \max_m [\alpha_{i-1}(j) a_{i-1,j,k} \hat{a}_{i,k,m} \hat{b}_{i,m}(\Delta\Theta_i)] b_{i,k}(M_i),$$

$$\Phi_i(k) = \arg\max_j\{\max_m[\alpha_{i-1}(j)a_{i-1,j,k}\hat{a}_{i,k,m}\hat{b}_{i,m}(\Delta\Theta_i)]\},$$
$$\hat{\Phi}_i(k) = \arg\max_m\{\max_j[\alpha_{i-1}(j)a_{i-1,j,k}\hat{a}_{i,k,m}\hat{b}_{i,m}(\Delta\Theta_i)]\},$$
$$2 \le i \le N-1.$$

Step 3: Termination.

$P^*(M, \Delta\Theta|\lambda) = \max_j[\alpha_{N-1}(s_j)]$.

The finial states are $s_{N-1,q_{N-1}}$ and $\hat{s}_{N-1,\hat{q}_{N-1}}$, where

$$q_{N-1} = \arg\max_j[\alpha_{N-1}(j)], \text{ and } \quad \hat{q}_{N-1} = \hat{\Phi}_{N-1}(q_{N-1}).$$

Step 4: State path backtracking.

The states at frequency i are s_{i,q_i} and $\hat{s}_{i,\hat{q}_i}$,

where $\quad q_i = \Phi_{i+1}(q_{i+1})$ and $\hat{q}_i = \hat{\Phi}_i(q_i)$.

The modified Viterbi algorithm not only gives the maximum of the output probability produced by a single-state path, but also discovers the optimal state path.

B: Re-estimation of model parameters

In practical applications, one single observation sequence (one training sample) is not enough for the re-estimation of the HMM's parameters. Let T denote the number of training samples. We can express the observations in the training set as $\mathbf{M}^T = \{M^1, M^2, \cdots, M^T\}$ and $\mathbf{\Delta\Theta}^T = \{\Delta\Theta^1, \Delta\Theta^2, \cdots, \Delta\Theta^T\}$, where $M^t = \{M_1^t, M_2^t, \cdots, M_{N-1}^t\}$ and $\Delta\Theta^t = \{\Delta\Theta_1^t, \Delta\Theta_2^t, \cdots, \Delta\Theta_{N-1}^t\}$. A very attractive property of the Viterbi algorithm is that it can retrieve the optimal state sequence, which makes the re-estimation procedure simple.

Let $s_{i,q_i^t}^t$ denote the state for the observation M_i^t, $n_{i,j}$ the total number of states $s_{i,q_i^t|q_i^t=j}^t$, $n_{i,j,k}$ the total number of state transitions from state $s_{i,j}^t$ to $s_{i+1,k}^t$, and $\hat{n}_{i,j,m}$ the total number of state transitions from $s_{i,j}^t$ to $\hat{s}_{i,m}^t$, where $1 \le t \le T$. Now the re-estimation formulas for π, $\mathbf{A}$ and $\hat{\mathbf{A}}$ are:

$$\pi_j = \text{the frequency in state } s_{1,j} = \frac{n_{1,j}}{T}, \tag{4.17}$$

$$a_{i,j,k} = \frac{\text{the number of state transitions from } s_{i,j}^t \text{ to } s_{i+1,k}^t}{\text{the number of state transitions from } s_{i,j}^t}$$
$$= \frac{n_{i,j,k}}{n_{i,j}}, \tag{4.18}$$

$$\hat{a}_{i,j,m} = \frac{\text{the number of state transitions from } s_{i,j}^t \text{ to } \hat{s}_{i,m}^t}{\text{the number of state transitions from } s_{i,j}^t}$$
$$= \frac{\hat{n}_{i,j,m}}{n_{i,j}}. \tag{4.19}$$

In order to simplify the re-estimation of $b_{i,j}(M_i)$ and $\hat{b}_{i,m}(\Delta\Theta_i)$, we assume that the codebook-tied output probability density functions take the form of the Gaussian function:

$$f(M_i^t, c_{i,v}) = G(M_i^t, \overline{M}_{i,v}, \sigma_{i,v}), \tag{4.20}$$

$$f(\Delta\Theta_i^t, \hat{c}_{i,v}) = G(\Delta\Theta_i^t, \overline{\Delta\Theta}_{i,v}, \hat{\sigma}_{i,v}), \tag{4.21}$$

where $G(\cdot,\cdot,\cdot)$ is the Gaussian probability density function, $\overline{M}_{i,v}$ and $\sigma_{i,v}$ are the mean and the variance of the codeword $c_{i,v}$, $\overline{\Delta\Theta}_{i,v}$ and $\hat{\sigma}_{i,v}$ are the mean and the variance of the codeword $\hat{c}_{i,v}$. These means and variances are pre-determined at the quantization stage. However, special care must be taken in calculating $\overline{\Delta\Theta}_{i,v}$ and $\hat{\sigma}_{i,v}$ due to the 2π wrap-around effect: for the nearest codeword of $\Delta\Theta_i^t$, the distance between the the codeword $\hat{c}_{i,v}$ and $\Delta\Theta_i^t$ is defined as $\min\{|\Delta\Theta_i^t - \overline{\Delta\Theta}_{i,v}|, 2\pi - |\Delta\Theta_i^t - \overline{\Delta\Theta}_{i,v}|\}$ to take the 2π wrap-around effect into consideration.

We define $Pr(c_{i,v}|s_{i,j})$ as the ratio of the expected number of times in state $s_{i,j}$ with the observation codeword $c_{i,v}$ to the number of times in state $s_{i,j}$. With the same token, we define $Pr(\hat{c}_{i,v}|\hat{s}_{i,j})$. Thus, we have

$$Pr(c_{i,v}|s_{i,j}) = \frac{\text{the number of times in state } s_{i,j} \text{ with observation codeword } c_{i,v}}{\text{the number of times in state } s_{i,j}}, \tag{4.22}$$

$$Pr(\hat{c}_{i,v}|\hat{s}_{i,j}) = \frac{\text{the number of times in state } \hat{s}_{i,j} \text{ with observation codeword } \hat{c}_{i,v}}{\text{the number of times in state } \hat{s}_{i,j}}. \tag{4.23}$$

C: Initial estimation of HMMs

In theory, the re-estimation procedure in HMMs should lead the initial model to a local optimum, because the optimization surface is usually complex with many local optima. It is well known that the initial considtions play a very important role in obtaining good models [111]. For example, the re-estimation of continuous probability density functions (pdf's) usually requires good, initial estimates [209]. In SCHMMs, this becomes relatively simple and reliable, because it uses the pre-determined, codebook-tied continuous pdf's. As a result, the number of free model parameters is the same as that of DHMMs. Accordingly, the method for model initialization in SCHMMs is similar to that used in DHMMs. We adopt the uniform initial estimates of π and state transition matrixes, which has shown to be adequate for the re-estimation of these parametes in almost all cases [209].

For the probabilities of the state codewords, we first divide V codewords into K subsets, then assign a constant ($> 1/V$ and $< K/V$) to the probabilities of the codewords in the jth ($j \leq K$) subset for state $s_{i,j}$ or $\hat{s}_{i,j}$ and a smaller constant ($< 1/V$) to the probabilities of the codewords in other subsets for this state.

4.4 Experimental Results

In unconstrained handwritten numerals, there are large variations in writing styles, e.g, cross-bars, ligatures, embellishments, filled holes and opening of loops, etc. Therefore, some numerals are of totally different shapes in different writing styles. It is a challenging task to use one model to represent the variations for one class or category. To evaluate the method presented in this chapter, we use the same digit database (goodbs) in CEDAR CDROM1 as that used in Chapter 3.

In the experiments, we trained the HMMs by the modified Viterbi algorithm discussed in Section 4.3 with three different numbers of states for the magnitude and the phase at each frequency. Sixteen codewords were generated for the magnitude and the phase at each frequency. Table 4.1 presents the correct recognition rates of HMMs with different number of states. The result shows clearly that the recognition accuracy is strongly dependent on the number of states. We can see that when an HMM has only two states ($K = 2$) for each part of the spectrum at a specific frequency, the recognition performance is reletively low due to its limited representational capability for shape variations in the handwritten numerals. When $K = 4$, however, the HMM was able to adequately represent shape variations, achieving a recognition rate of 96.6%. Further increasing the number of states did not contribute to a significant increase in recognition rate; for instance, when $K = 8$, the recognition rate increased slightly to 96.7%.

TABLE 4.1. The performance of the SCHMM with spectral features.

state number K	2	4	8
Recognition rates	93.3%	96.6%	96.7%

It is interesting to compare the performance of this method with that of other similar HMM-based methods. We have compared four Markov model-based methods: 1-D HMM[1] [190], 1-D HMM[2] [80], 1-D HMM[3] [37], pseudo 2-D HMMs [158] and 2-D HMMs [189], where three 1-D HMM-based methods use different features. Park and Lee used four projected histograms as features to form four 1-D HMMs [190]. Elms extracted features from rows and columns by Fourier transformation and quantized the spectra into codewords, from which they obtained the row/column DHMMs for handwriting recognition [80]. Cai and Liu extracted features from outer contours of images and arranged them into sequences for 1-D HMMs [37]. Table 4.2 shows the performance of these HMM-based approaches.

TABLE 4.2. The performance comparison of HMM-based methods.

Methods	Number of states	Recognition rates
1-D HMM[1]	8	82.0%
1-D HMM[2]	10	95.6%
1-D HMM[3]	varied	98.0%
Pseudo 2-D HMM	10	85.2%
2-D HMM	4	91.6%

From Tables 4.1 and 4.2, we can see clearly that the recognition rate of the new method ($K = 4$) is much higher than or comparable to those published recently. It is also clear from the results of the proposed method and 1-D HMM[3] that features extracted from outer contours are suitable for 2-D shape recognition.

4.5 Discussion

From the comparison, we see a well-known phenomenon in pattern recognition that features play a significant role in the performance of recognizers. With 1-D HMMs[1], for similar characters, we cannot obtain features that have good discriminant characteristics [190], which often results in poor recognition rates. For the methods based on 2-D HMMs [189] and pseudo 2-D HMMs [158], features used in these systems are extracted from mesh regions. However, since handwritten characters are composed of elongated patterns, features extracted from such regions are not suitable for representing elongated patterns. Moreover, minor slant or rotation of images may produce significant changes in such features. The major improvement of 2-D HMMs over the pseudo 2-D HMMs is that a third-order causal neighborhood system, which is an extension of the first-order neighbour system in 1-D HMM[1] and pseudo 2-D HMMs, is applied to exploiting the spatial correlation.

In this chapter, we have presented a Markov model for modeling shape features in the spectral space. We have analyzed the properties of the contour-based Fourier spectra and presented methods for extracting sequential features, which are robust to slant and rotation within the range ($0°$-$30°$), and modeling these features using HMMs. We have also modified the conventional Viterbi algorithm for re-estimating model parameters. We must stress that this model can be extended to the recognition of general 2-D patterns with variations in orientation and reflection, etc. The experimental results demonstrate that the features extracted from outer contour

Fourier transformation are suitable for representing deformable shapes in handwritings, and that the recognition accuracy of this method is at least comparable to the most recently published results.

5

Markov Random Field Model for Recognizing Handwritten Digits

Markov random field (MRF) models are multidimensional in nature that allows us to combine both statistical and structural information for pattern recognition. Therefore, MRF models have been widely applied to image restoration, segmentation, texture modeling and classification.

In this chapter, we use MRF models to recognize handwritten numerals. We briefly introduce the theory of MRF models. We then present a MRF-based method for recognizing handwriting numerals and discuss some important issues on using MRF models including global similarity and the design of neighborhood and clique functions.

5.1 Fundamentals of Markov Random Fields

This section introduces the fundamental theory of Markov random field models as well as background materials that will be used in this and next chapters. As MRF models are the natural extension of 1-D Markov models, this section begins with 1-D Markov processes.

5.1.1 One-Dimensional Markov Processes

Let us first introduce one-dimensional unilateral Markov processes. Let X be a random process, where $X = \{X_k, k \in K\}$ and $K = \{0, \pm1, \pm2, \cdots\}$. If X is an Nth order causal Markov process, then we have [70]

$$Pr(X_k|X_{k-1}, \cdots, X_{k-M}) = Pr(X_k|X_{k-1}, \cdots, X_{k-N}), \qquad (5.1)$$

where $M \geq N$. Papoulis [191] derived the "chain rule" property for general Nth order unilateral Markov processes:

$$\begin{aligned} Pr(X_k, \cdots, X_{k-n}) = \; & Pr(X_k|X_{k-1}, \cdots, X_{k-N}) Pr(X_{k-1}|X_{k-2}, \cdots, X_{k-N-1}) \\ & \cdots Pr(X_{k-n+N}|X_{k-n+N-1}, \cdots, X_{k-n}) \\ & Pr(X_{k-n+N-1}, \cdots, X_{k-n}), \end{aligned} \qquad (5.2)$$

where $n > N$. It is very interesting to note the special case when $N = 1$ and $K = \{1, 2, 3, \cdots\}$:

$$Pr(X_k|X_1, X_2, \cdots, X_{k-1}) = Pr(X_k|X_{k-1}). \qquad (5.3)$$

From (5.3), we can obtain a simple and useful property:

$$Pr(X_1, X_2, \cdots, X_k) = Pr(X_1)Pr(X_2|X_1)\cdots Pr(X_k|X_{k-1}), \tag{5.4}$$

where $Pr(X_1)$ is the boundary conditional probability. This property allows the probability of X to be represented by a product of all conditional probabilities. As a result, $Pr(X_1, X_2, \cdots, X_k)$ over all possible X or the optimal path can be calculated in a lattice by using some efficient algorithms such as the forward algorithm [11] and the Viterbi algorithm [249] in HMMs, where X is a sequence of states.

X is an Nth order bilateral Markov process if and only if:

$$Pr(X_k|X_{k-I}, \cdots, X_{k+M}) = Pr(X_k|X_{k-N}, \cdots, X_{k+N}) \tag{5.5}$$

holds for any N $(I, J \geq N)$.

The theory of MRFs was initially proposed by Ising and Lenz based on the concept of the Nth order bilateral Markov processes [29, 140]. Ising and Lenz established a general probabilistic model (now called the *Ising model*) to explain certain empirically observed facts about ferromagnetic materials. Let us consider a sequence of M sites in a line. At each site, there is a spin ω_i which takes one of two forms, '+' and '-', representing the property of the ferromagnetic materials. Thus the spins in the form of a configuration, $\omega = \{\omega_1, \omega_2, \cdots, \omega_M\}$, can be shown as in Figure 5.1.

$$+ + - - - + + - + - - + +$$

FIGURE 5.1. The configuration of spins in the Ising model

Ising [140] defined a function $f(\omega_i)$ on the Ω space so that $f(\omega_i) = 1$ if $\omega_i =$ '+' and $f(\omega_i) = -1$ if $\omega_i =$ '-'. Ising also defined a probability measure on the Ω space based on the energy function $U(\omega)$. For a given configuration ω, the energy $U(\omega)$ is assigned by [140]

$$U(\omega) = -J\sum_i\sum_j f(\omega_i)f(\omega_j) - mH\sum_i f(\omega_i), \tag{5.6}$$

where J and m are constants determined by the properties of materials and H is the intensity of an external magnetic field. And the probability of the configuration ω is given by

$$Pr(\omega) = \frac{1}{Z} exp\left\{-\frac{1}{k_0 T}U(\omega)\right\}, \tag{5.7}$$

where the normalcies constant Z is

$$Z = \sum_\omega exp\left\{-\frac{1}{k_0 T}U(\omega)\right\}, \tag{5.8}$$

which is also called the partition function. In order to simplify the model, Ising considered only the interactions between neighboring spins (the first order Markov process). Thus, the local energy functions and the probability of the configuration for the Ising model can be given by

$$U_i(\omega) = -\frac{J}{2} \sum_{|j-i|=1} f(\omega_i) f(\omega_j) - mHf(\omega_i), \tag{5.9}$$

$$Pr(\omega) = \frac{1}{Z} \prod_i exp\left\{ -\frac{1}{k_0 T} U_i(\omega) \right\}. \tag{5.10}$$

As a result, the probability of the configuration can be obtained from local information. For 2-D configuration in a lattice as shown in Figure 5.2, the Ising model can handle the interaction between spins provided that the neighborhood system is well defined.

$$\begin{array}{ccccccccccccc}
+ & + & - & - & - & + & + & - & + & - & - & + & + \\
+ & - & + & + & - & - & - & + & + & - & + & + & - \\
- & - & + & - & + & - & + & - & + & + & - & - & + \\
- & - & - & + & - & + & + & - & - & + & + & + & - \\
- & + & - & + & - & + & - & - & + & - & + & + & + \\
+ & + & - & + & + & - & - & - & + & + & - & - & +
\end{array}$$

FIGURE 5.2. The 2-D configuration of spins in Ising model

5.1.2 *Markov Random Fields*

In recent years, MRF models have gained growing popularity in various applications, such as image restoration [14, 94, 97, 128], texture classification [47, 188, 232], edge detection [95, 258], speech recognition [226], image interpretation [179], etc. The popularity of modeling images via local interactions of random variables specified by MRFs is due to the following main reasons [163]:

- MRFs can be used to develop algorithms based on a sound theory in statistics instead of heuristics.
- Modeling images based on local interactions between random variables allows the algorithms to be implemented in a parallel fashion.
- The theory of MRFs provides a basis for modeling contextual constraints by the Gibbs distribution, so that the prior information can be easily incorporated quantitatively.
- MRFs are able to model structural information as well as statistical information.

Consider a random field X defined on lattice L, where L is a discrete rectangular lattice $L = \{i, j\}$.

Definition 5.1.1
X is a Markov random field with respect to $(L, \mathcal{N})$ if and only if

$$Pr(X_{ij}|X_{kl}, (k,l) \in \Omega) = Pr(X_{ij}|X_{kl}, (k,l) \in \mathcal{N}_{ij}), \tag{5.11}$$

where $X_{ij} \in X$ is a random variable at site (i, j), $\mathcal{N}_{ij}$ is the neighborhood of (i, j), and Ω is any finite subset of the lattice L that contains N_{ij} but not (i, j).

The neighborhood system associated with the lattice L is designed as follows [70]:

$$\mathcal{N} = \{\mathcal{N}_{ij} \subset L; (i,j) \in L\}, \tag{5.12}$$

where the neighborhood of (i, j), $\mathcal{N}_{ij}$, has following property:

$$(i,j) \notin \mathcal{N}_{ij}. \tag{5.13}$$

If the neighborhood system is homogeneous, there is

$$(i,j) \in \mathcal{N}_{kl}, \qquad \text{if } (k,l) \in \mathcal{N}_{ij}. \tag{5.14}$$

A: Gibbs Distributions

The MRF is defined on a specific neighborhood system on a lattice. The Hammersley-Clifford theorem states that X is an MRF with respect to the $\mathcal{N}$ and $Pr(X = \omega) > 0$ for all ω if and only if X is a Gibbs random field with respect to $\mathcal{N}$ and the associated cliques [233, 13]. That is, in the neighborhood system $\mathcal{N}$, the MRF can be modeled by a Gibbs (Boltzmann) distribution. Besag showed that under a very mild condition in MRF the prior distribution function is Gibbsian [13]:

$$Pr(X = \omega) = \frac{1}{Z}\exp\{-U(\omega)\} = \frac{1}{Z}\exp\{-\sum_{c \in C} V_c(\omega)\}, \tag{5.15}$$

where Z is the normalizing constant and $U(\omega)$ is the energy function as that of the Ising model, C is the set of all cliques, c is a clique and $V_c(\omega)$ is a potential assigned to clique c, where $c \in C$. $V_c(\omega)$ is also called the *clique* function. Note that there is no special requirement on the design of neighborhood systems and clique functions as long as the clique functions depend only on the sites in the corresponding cliques. Consequently, different neighborhood systems and clique functions can be used to accommodate varied applications in the framework of MRFs.

B: Neighborhood System

MRFs are defined on neighborhood systems, and different neighborhood systems have different properties and serve different purposes. Homogeneous neighborhood systems are widely used in image restoration. Geman and Geman used homogeneous neighborhood systems of the form $\mathcal{N} = \{\mathcal{N}_{ij}\}$, where

$$\mathcal{N}_{ij} = \{(k,l); 0 < (k-i)^2 + (l-j)^2 \leq d\}, \tag{5.16}$$

in which d is a constant [94].

Figure 5.3 gives examples of nearest-neighborhood configurations and the cliques used in [94, 70]. Figures 5.3(a), (b) and (c) show the neighborhood configurations for $d = 1, 2, 4$. The neighborhood system shown in Figure 5.3(a) is first order, in which the neighborhood of the site (i, j) is $\mathcal{N}_{ij} = \{(i, j-1), (i, j+1), (i+1, j), (i-1, j)\}$. The cliques of the first-order neighborhood system are depicted in Figure 5.3(d). The second-order neighborhood system is shown in Figure 5.3(b)) and its cliques are depicted in Figures 5.3(d) and (e).

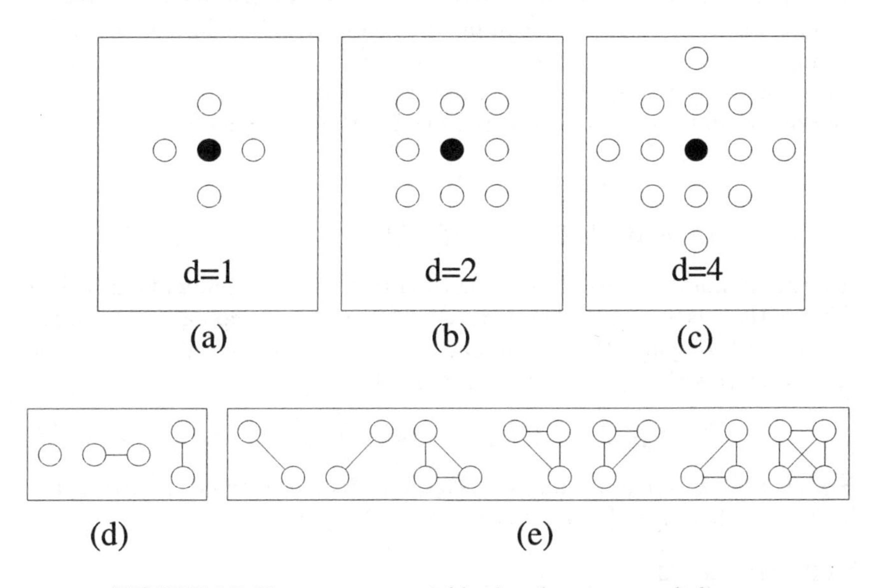

FIGURE 5.3. Homogeneous neighborhood systems and cliques.

Geman and Geman considered line processes in designing their clique functions for image segmetation applications [94]. Different potentials V_c were assigned to cliques with different conditions. Figure 5.4 shows the potentials of cliques with six distinct types of rotations used in [94].

In edge detection, the labeling problem is to assign a label to a site from the set {Edge, Non-edge}. In conventional methods, edges are marked

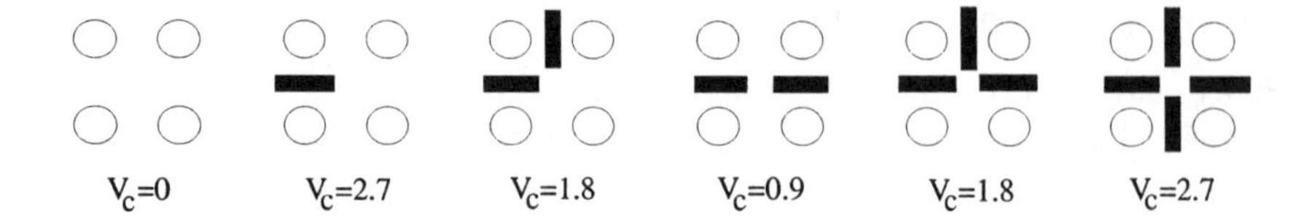

FIGURE 5.4. Potentials assigned to different types of cliques.

at local maximums of the image convoluted with a first derivative operator [214, 216] or at the location where the second derivative of the image crosses zero intensity values [174, 105]. However, these methods are sensitive to noise. Canny located edges by using the optimal operator based on signal to noise ratio considerations [42]. Canny's method is robust to noise but it has two problems: 1) biases in edge locations, and 2) broken edges at corners and crosses. Ulupinar and Medioni proposed a method to reduce effects of the first problem, so that they could use a post-processing to deal with the second problem [245].

For edge labeling using MRFs, most researchers chose the clique potentials that tend to support the continuity of line segments and discourage abrupt breaks and sharp turns in line segments and close parallel lines. For example, Chou *et al.* assigned a negative value to a clique with line continuity and a positive value to a clique with line end or parallel lines [62]. Figure 5.5 and Figure 5.6 show the neighborhood system and the cliques used in [62].

5.1.3 Markov Mesh Random Fields

Another important type of MRFs is the Markov mesh random fields (MMRFs) with causal neighborhood systems. Since there is no natural ordering of sites in 2-D space, there is no natural definition of causality. We define the past Ψ_{ij} with respect to the current site (i,j) as

$$\Psi_{ij} = \{(k,l) : 1 \le k \le i,\ 1 \le l \le j,\ (k,l) \ne (i,j)\}. \tag{5.17}$$

Now, consider a random field, $X = \{X_{ij}\}$, defined over an $m \times n$ rectangular lattice L, where $L = \{(i,j) : 1 \le i \le m,\ 1 \le j \le n\}$.

Definition 5.1.2
X is a Markov mesh random field (MMRF) if and only if:

$$Pr(X_{ij}|X_{kl} \in \Phi_{ij} = Pr(X_{ij}|X_{kl} \in \Lambda_{ij}) \tag{5.18}$$

for all $(i,j) \in L$, *where* $\Phi_{ij} = \{(k,l) : k < i \text{ or } l < j\}$, *and* $\Lambda_{ij} \subset \Psi_{ij}$ *is the support set of the site* (i,j).

The relationships among sets Ψ_{ij}, Φ_{ij} and Λ_{ij} of the site (i,j) associated with MMRF are depicted in Figure 5.7.

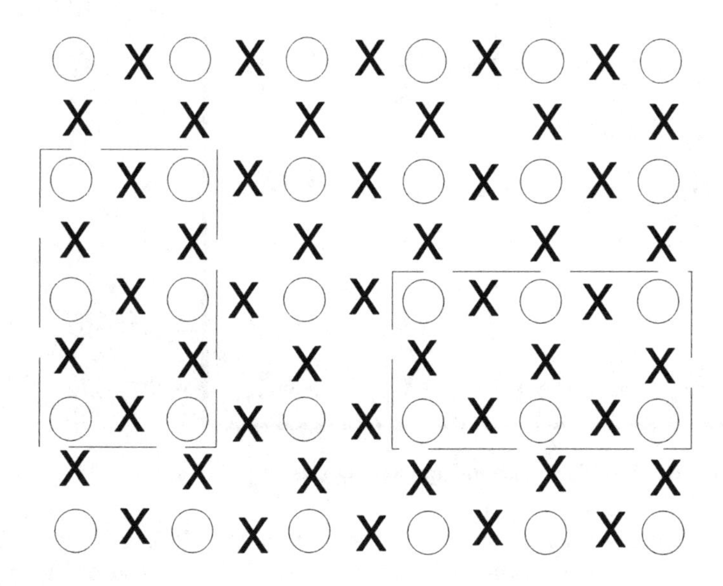

FIGURE 5.5. Edge sites in the second neighborhood system, where Xs indicate line (edge) and circles represent pixels.

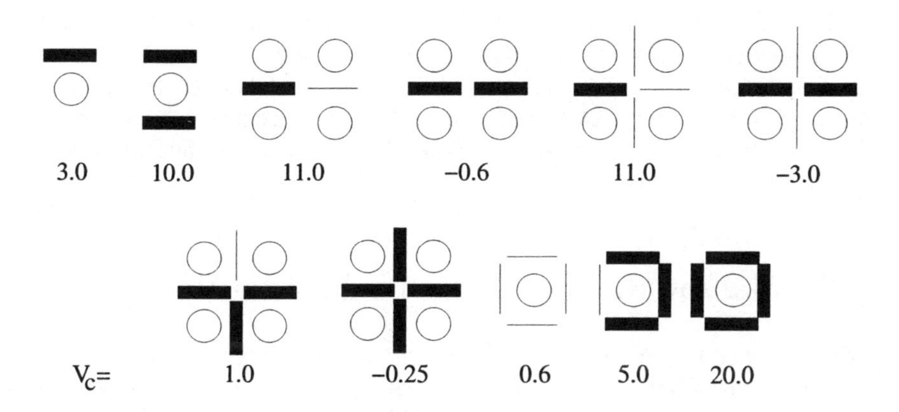

FIGURE 5.6. Potentials assigned to different cliques are used for corrupted edge data, where edges and non-edges are represented by solid and dotted lines, respectively.

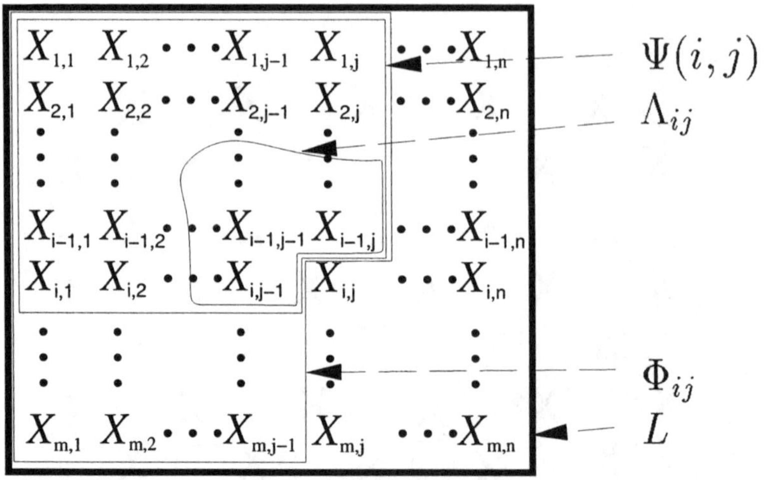

FIGURE 5.7. The relationships among sets Ψ_{ij}, Φ_{ij} and Λ_{ij} of the (i,j).

The support set Λ_{ij} can be any subset of Ψ_{ij}. It is not necessary to have a fixed configuration with respect to (i,j). For instance, the support set, which is used in [39] and the previous chapter, consists of only one site: $\Lambda_{ij} = \{(i-1,j)\}$ for the phase spectrum and $\Lambda_{ij} = \{(i,j-1)\}$ for the magnitude spectrum. However, the support set usually has a fixed configuration with respect to (i,j). The commonly used support set is:

$$\Lambda_{ij} = \{(i,j-1),(i-1,j),(i-1,j-1)\}, \tag{5.19}$$

which was used in 2-D HMMs [189]. For such causal support sets [39, 189], the joint distribution of $\{X_{ij},(i,j) \in L\}$ can be expressed as the product of causal conditional distributions. This property of MMRF is analogous to the similar property of 1-D first-order Markov processes. Therefore, the 2-D Viterbi algorithm can be used to derive the global optimum from local conditional distributions.

5.2 Markov Random Field for Pattern Recognition

MRFs were not familiar to most researchers in image processing and pattern recognition until the pioneering work of Geman and Geman [94], Besag [13, 14] and Kindermann and Snell [140]. Since the mid 1980s, the theory of MRFs has found many applications in low-level image processing. MRFs can be used to model general image structures such as interactions between neighboring pixels, line-segments or regions. Moreover, MRFs can be defined not only over regular sites but also irregular sites. Recently MRFs have been used in some high-level signal processing recognition applications, such as solid object recognition [159, 160] and speech recognition [118].

In this section, we will present an MRF-based approach to general pattern recognition tasks as well as to handwritten numeral recognition.

5.2.1 *Maximum* a posteriori *Probability*

The use of statistical or Bayesian paradigms for pattern recognition has been well known [163]. In the Bayesian framework, the recognition requires an estimation process that maximizes *a posteriori* (MAP) probability introduced in Chapter 3. Let us recall the concept of maximum *a posteriori* (MAP) probability. Given observation O and class model λ_w, we may express the conditional probability that a pattern belongs to class w by (3.5) which we rewritten here for convenience:

$$Pr(w|\mathbf{O}, \lambda_w) = Pr(w)\frac{Pr(\mathbf{O}|\lambda_w)}{Pr(\mathbf{O})}, \tag{5.20}$$

Suppose that the model λ_w has N states, then $Pr(\mathbf{O}|\lambda_w)$ can be expressed as

$$Pr(\mathbf{O}|\lambda_w) = \sum_{\text{all } \mathbf{X}} Pr(\mathbf{O}, \mathbf{X}|\lambda_w), \tag{5.21}$$

where $\mathbf{X}$ is a set of states,

$$Pr(\mathbf{O}, \mathbf{X}|\lambda_w) = Pr(\mathbf{O}|\mathbf{X}, \lambda_w)Pr(\mathbf{X}|\lambda_w), \tag{5.22}$$

$Pr(\mathbf{X}|\lambda_w) = Pr(X_1 = q_1, X_2 = q_2, \cdots, X_i = q_i, \cdots, X_I = q_I)$ and I is the number of observations or short segments. Unfortunately, as discussed in Section 3.1.3 the amount of direct computation of (5.21) is unfeasible even in one-dimensional space. There are two major techniques: the forward-backward algorithm [209] and the Viterbi algorithm [249]. For 2-D cases, if the neighborhood system is causal, we may use the evaluation techniques proposed in [189, 39, 70]. In general, however, the locations of short segments are irregular, it is impossible to defined a causal neighborhood system on irregular sites. Therefore, these techniques are not readily applicable to the cases that involve multi-dimensional irregular sites. Instead, we may compute $Pr(\mathbf{O}|\lambda_w)$ by approximation (similar to that in the Viterbi algorithm [69]):

$$Pr(\mathbf{O}|\lambda_w) \approx Pr(\mathbf{O}, \mathbf{X}_{best}|\lambda_w), \tag{5.23}$$

where

$$\mathbf{X}_{best} = \arg\max_{\mathbf{X}}\{Pr(\mathbf{O}, \mathbf{X}|\lambda_w)\}. \tag{5.24}$$

Thus, the problem of recognition becomes that of obtaining $Pr(\mathbf{X}_{best}|\lambda_w) \times Pr(\mathbf{O}|\mathbf{X}_{best}, \lambda_w)$ and the best state set $\mathbf{X}_{best}$ [4]. In other words, the problem of evaluation is to label the best states in terms of maximizing $Pr(\mathbf{X}|\lambda_w)\ Pr(\mathbf{O}|\mathbf{X}, \lambda_w)$.

5.2.2 Markov Random Fields for Modeling Statistical and Structural Information

The problem of best state estimation is equivalent to the problem of optimal labeling. A short segment i is optimally labeled as state X_i means that the state X_i can optimally interpret the observation o_i. If the interaction between states exists only within a certain region, we can regard the labeling configuration as an MRF. Let $\mathbf{X}$ be an MRF on $\mathbf{L}$, where $\mathbf{L}$ can be a lattice with regular or irregular sites. Let $\mathcal{N}$ be the neighborhood system on $\mathbf{L}$

$$\mathcal{N} = \{\mathcal{N}_i | \forall i \in I\}, \tag{5.25}$$

where $\mathcal{N}_i$ is the collection of observations (or short segments) that are the neighbors of the observation denoted by o_i. A clique c defined on $(\mathbf{X}, \mathcal{N})$ can be a group sites or a singleton. $\mathbf{X}$ is an MRF on $\mathbf{L}$ with respect to a homogeneous neighborhood system $\mathcal{N}$ if and only if

- $Pr(\mathbf{X}|\lambda_w) > 0, \qquad \forall \mathbf{X};$
- $Pr(X_i|X_j, \lambda_w;\ i \neq j\ \&\ \forall i, j \in I) = Pr(X_i|X_j, \lambda_w; j \in \mathcal{N}_i),$

where $Pr(\mathbf{X}|\lambda_\mathbf{w})$ is the joint probability distribution function for the given model λ_w and $Pr(X_i|X_j, \lambda_w)$ is the conditional probability distribution function. According to the Hammersley-Clifford theorem, the joint probability distribution functions of MRFs with respect to the neighborhood system $\mathcal{N}$ have a general functional form, known as the Gibbs distribution, which is defined based on the concept of cliques [13]. Therefore, the joint probability distribution function $Pr(\mathbf{X}|\lambda_\mathbf{w})$ can be expressed in terms of the Gibbsian form:

$$p(\mathbf{X}|\lambda_w) = Z^{-1} e^{-U_1(\mathbf{X}|\lambda_w)}, \tag{5.26}$$

where

$$U_1(\mathbf{X}|\lambda_w) = \sum_{c \in \mathbf{C}} V_{c1}(\mathbf{X}|\lambda_w) \tag{5.27}$$

is the energy function and $V_{c1}(\cdot)$ the clique function defined on the corresponding clique $c \in \mathbf{C}(\mathbf{L}, \mathcal{N})$, where $\mathbf{C}(\mathbf{L}, \mathcal{N})$ is the collection of all the cliques.

Let us take $Pr(\mathbf{O}|\mathbf{X}, \lambda_w)$ into consideration. Substituting (5.26) into (5.22), we have

$$Pr(\mathbf{O}, \mathbf{X}|\lambda_w) = Z^{-1} e^{-U(\mathbf{O}, \mathbf{X}, \lambda_w)}, \tag{5.28}$$

where

$$U(\mathbf{O}, \mathbf{X}, \lambda_w) = U_1(\mathbf{X}|\lambda_w) + U_2(\mathbf{O}|\mathbf{X}, \lambda_w) \tag{5.29}$$

is the global likelihood energy function and

$$U_2(\mathbf{O}|\mathbf{X}, \lambda_w) = -\log Pr(\mathbf{O}|\mathbf{X}, \lambda_w).$$

Now, the problem of labeling optimal states is equivalent to minimizing the global likelihood energy function $U(\mathbf{O}, \mathbf{X}, \lambda_w)$. In order to simplify the task of finding the global energy minimum, we assume that the observations $\{o_1, \cdots, o_I\}$ are conditionally independent, which enables us to express $Pr(\mathbf{O}|\mathbf{X}, \lambda_w)$ as follows:

$$Pr(\mathbf{O}|\mathbf{X}, \lambda_w) = \prod_{i=1}^{I} Pr(o_i|X_i, \lambda_w). \tag{5.30}$$

Consequently,

$$U_2(\mathbf{O}|\mathbf{X}, \lambda_w) = \sum_{c \in \mathbf{C}} V_{c2}(\mathbf{O}|\mathbf{X}, \lambda_w) = \sum_{i=1}^{I} - \log Pr(o_i|X_i, \lambda_w), \tag{5.31}$$

and

$$U(\mathbf{O}, \mathbf{X}, \lambda_w) = \sum_{c \in \mathbf{C}} \{ \underbrace{V_{c1}(\mathbf{X}|\lambda_w)}_{structural} + \underbrace{V_{c2}(\mathbf{O}|\mathbf{X}, \lambda_w)}_{statistical} \}, \tag{5.32}$$

where $V_{c1}(\mathbf{X}|\lambda_w)$ is used to model the local structural information that is described by the relationships among states within the neighborhood system, and $V_{c2}(\mathbf{O}|\mathbf{X}, \lambda_w)$ models the statistical information because it is a logarithmic function of the output probability for the given observation and state. In this framework, the task of finding the best set of states in the recognition process is now equivalent to minimizing the global likelihood energy function in (5.32). There are several algorithms that can be used to locate the global optimal point, e.g., simulated annealing [141] and relaxation labeling [144]. Therefore, MRFs are ready to be used to model both structural and statistical information for pattern recognition.

5.2.3 Neighborhood System and Cliques

We will discuss how to design a neighborhood system and cliques for an MRF. Because the recognition process is to minimize the likelihood energy function that is the summation of the clique functions defined on the neighborhood system, the design of the neighborhood system and cliques is critical to the success of MRF-based systems. Different definitions of neighborhood systems and cliques, which result in various types of MRF models, are used to solve different problems. For most applications, the neighborhood systems are defined on regular lattices [70].

For high-level vision applications, neighborhood systems are usually defined on irregular sites. Although neighborhood systems can be of any shapes and sizes in theory, the design of neighborhood systems and cliques is based on connection and distance.

Figure 5.8 shows a neighborhood system and cliques based on connection on irregular sites. Some patterns are represented by feature points, and many feature points are not connected directly with others. Therefore, we adopt the neighborhood system that consists of the sites within a certain distance. Figure 5.9 shows a neighborhood system and cliques based on distances. This kind of neighborhood systems is particularly suitable for recognizing patterns or retrieving images by critical point or feature point matching [223], because not all critical points have to be connected.

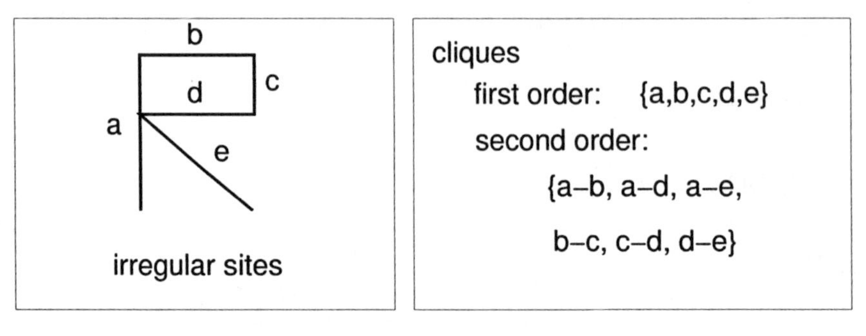

FIGURE 5.8. The neighborhood system and cliques based on connectedness.

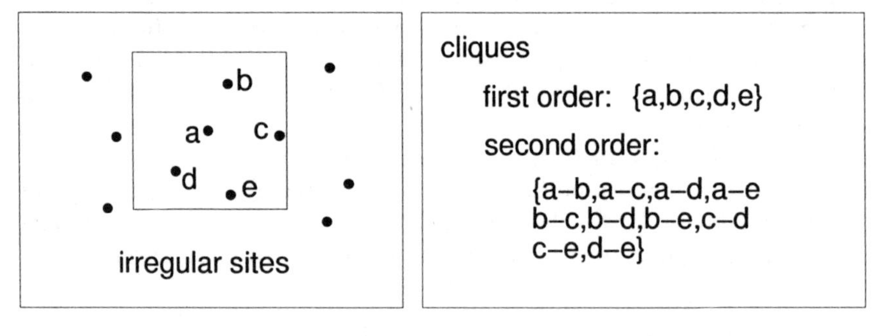

FIGURE 5.9. The neighborhood system and cliques based on distance.

5.2.4 *Minimizing the Likelihood Energy*

A: Viterbi Algorithm

If the neighborhood system of an MRF is causal, we can use the Viterbi algorithm to solve optimization problems. Suppose that the global likelihood energy is given by

$$\begin{aligned} U(\mathbf{O}, \mathbf{X}, \lambda_w) &= V_{c1}(X_1|\lambda_w) + V_{c2}(o_1|X_1, \lambda_w) + V_{c1}(X_2|X_1, \lambda_w) \\ &+ V_{c2}(o_2|X_2, \lambda_w) + \cdots + V_{c1}(X_I|X_{I-1}, \lambda_w) + V_{c2}(o_I|X_I, \lambda_w). \end{aligned} \tag{5.33}$$

We need to find the minimum energy passing the site i which is dependent on the clique functions for the site i and the minimum energy passing the previous site for the given state labels. For such cases, we may implement the Viterbi algorithm by the following steps.

Initially, the minimum energy passing the first site

$$E_1(X_1) = V_{c1}(X_1|\lambda_w) + V_{c2}(o_1|X_1, \lambda_w). \tag{5.34}$$

The minimum energy passing the second site and other sites can be expressed in turn as follows:

$$E_2(X_2) = \min_{X_1}\{E_1(X_1) + V_{c1}(X_2|X_1, \lambda_w) + V_{c2}(o_2|X_2, \lambda_w)\} \tag{5.35}$$

$$\vdots$$

$$E_i(X_i) = \min_{X_{i-1}}\{E_{i-1}(X_{i-1}) + V_{c1}(X_i|X_{i-1}, \lambda_w) + V_{c2}(o_i|X_i, \lambda_w)\} \tag{5.36}$$

$$\vdots$$

Therefore, the minimum of the global likelihood energy can be obtained by

$$U(\mathbf{O}, \mathbf{X}_{optimal}, \lambda_w) = \min_{X_I}\{E_I(X_I)\}. \tag{5.37}$$

However, it is difficult (if at all possible) to define a suitable causal neighborhood system over multidimensional irregular sites. Therefore, for irregular sites the Viterbi algorithm is not applicable in general.

B: Iterated Conditional Modes

Since it is difficulty to obtain the global minimum of the likelihood energy of an MRF, Kittler and Föglein proposed a deterministic algorithm based on local dependence [142], called iterated conditional modes (ICM), which is computationally graceful due to its appealing local characteristics. As $\mathbf{X}$ is an MRF, the state X_i depends only on the state labels in the local neighborhood:

$$Pr(X_i|X_k, \lambda_w; k \neq i) = Pr(X_i|X_k, \lambda_w; k \in \mathcal{N}_i). \tag{5.38}$$

Therefore, the likelihood energy in (5.32) can be expressed as

$$U(\mathbf{O}, \mathbf{X}, \lambda_w) = \sum_i U_i, \tag{5.39}$$

where

$$U_i = \{V_{c2}(o_i|X_i, \lambda_w) + \sum_{c_i \in \mathbf{C}_i} V_{c1}(X_i|X_k, \lambda_w; k \in \mathcal{N}_i)\}, \tag{5.40}$$

c_i is a clique in $\mathcal{N}_i$ and $\mathbf{C}_i$ is the collection of all cliques in $\mathcal{N}_i$. Obviously, it is much easier to minimize U_i than the global likelihood energy. This is the basis of ICM. In the ICM scheme, it is assumed that the states in $\mathcal{N}_i$ are known, so that U_i can be minimized by

$$U_i^{(t+1)} = \min_{X_i}\{V_{c2}(o_i|X_i, \lambda_w) + \sum_{c_i \in \mathbf{C}_i} V_{c1}(X_i|X_k^{(t)}, \lambda_w; k \in \mathcal{N}_i)\}, \quad (5.41)$$

and the estimate of state X_i can be obtained by

$$X_i^{(t+1)} = \arg\min_{X_i}\{V_{c2}(o_i|X_i, \lambda_w) + \sum_{c_i \in \mathbf{C}_i} V_{c1}(X_i|X_k^{(t)}, \lambda_w; k \in \mathcal{N}_i)\}, \quad (5.42)$$

where $U_i^{(t)}$ and $X_i^{(t)}$ denote the local energy and the state X_i after tth iterations, respectively. It is not difficult to find that the global likelihood energy never increases at any stage of ICM and the algorithm guarantees the eventual convergence. The experiments showed that the ICM algorithm converges to a local minimum of $U(\mathbf{O}, \mathbf{X}, \lambda_w)$ extremely rapid [14]. In addition, the ICM can be implemented by parallel computing techniques.

However, the performance of this algorithm is heavily dependent on the initial guess of $\mathbf{X}$. To deal with this problem, Besag proposed a simple modification to the basic version of ICM [14]. Besag used a sequence of weaker fields to update the labels of random variables on the first few iterations. This can prevent the pre-mature convergence on the basis of the unreliable initial estimate. Besag also suggested to replace a single site optimization in (5.41) by maximizing the joint probability over a small set X_B [14]. In this modification, $\mathbf{X}$ is supposed to be a Markov-P process [70]:

$$Pr(X_B|X_k, \lambda_w; k \notin X_B) = Pr(X_B|X_{\partial B}, \lambda_w), \quad (5.43)$$

where ∂B denotes the neighbors of the set B. In order to maintain the efficiency of ICM, B usually consists of only a few sites. However, in terms of solution quality, the ICM algorithm is not promising as it uses the "greedy" strategy in minimizing the local energy.

C: Simulated Annealing

Simulated annealing (SA), proposed by Kirkpatrick *et al.* [141] and Cerny [46], is a stochastic algorithm for combinatorial optimization. The impetus behind the SA algorithm is its simulation of the statistical mechanics of the physical annealing process, which cools down a melted compound slowly so that atoms are able to adopt the optimal configuration with low ensemble energy. However, if the cooling process is performed too quickly, the atoms will be trapped in non-optimal conformation with extensive irregularities. Consequently, the ensemble energy of the physical system is high. The simulated annealing adopts the similar procedure computationally. In a physical system, the probability of an atom at any energy level can be computed by using the Boltzmann distribution. Metropolis *et al.* developed an algorithm to simulate the behaviour of atoms in thermal equilibrium at a particular temperature with an initial configuration having the ensemble energy E_0 [177]. At each iteration, a small perturbation is given to the current configuration ω_i to generate the next configuration ω_j with the ensemble energy E_j. If $E_j \leq E_i$, the configuration ω_j is accepted. If $E_j > E_i$, the configuration ω_j may be accepted with a probability $\exp[-(E_j - E_i)/kT]$, where T is the temperature and k is a constant. The acceptance of bad moves based on the temperature T and the energy difference $E_j - E_i$ helps the simulated annealing algorithm to avoid being trapped in local optima. If the configuration ω_j is rejected, a configuration will be randomly selected as the current configuration and the above process will be repeated. In some systems, the Gibbs sampler is used to generate the next configuration based on a conditional probability instead of the ensemble energy change [94]. After a large number of iterations, the configurations reaches the equilibrium at the fixed temperature T. The next step is to reduce the temperature T and repeat the whole process again. In this way, the simulated annealing algorithm can guarantee to locate the global energy minimum if the cooling process is slowly *enough.*

How slowly must the temperature decrease? The answer to this question is application dependent. There is evidence that if the decreasing sequence satisfies [159]:

$$\lim_{t \to \infty} T^{(t)} = 0, \tag{5.44}$$

and

$$T^{(t)} \geq \frac{m \times \Delta}{\ln(1+t)}, \tag{5.45}$$

where t is number of iterations, m is a constant and $\Delta = \max_i E_i - \min_i E_i$, the system will converge to the global optimum regardless of the initial configuration ω_0. There are several other schedules for decreasing temperature for speeding up the algorithm and maintaining the convergence property [94, 146]. However, the SA algorithm is too slow to many applications.

D: Relaxation Labeling

Relaxation labeling is a process to assign labels to objects, feature points or components in a scene. Originally, relaxation labeling was developed to deal with ambiguity and noise [121]. Recently it has been applied to many problems in computer vision, ranging from edge detection [106] to scene interpretation [94] on the basis of labeled objects or components in the scene. However, relaxation labeling has much broader applications. The relaxation labeling algorithms are mainly based on two facts: 1) the problem of labeling will be much simpler if the complex computation can be decomposed into a network of local computations; and 2) the contextual information is useful in resolving ambiguities in computer vision. It is well known that MRF models provide flexible neighborhood systems for modeling the interactions between spatially related random variables. The global energy function of MRF is represented in terms of local energy functions. We can use relaxation labeling to combine contextual constraints in the neighborhood of the object or feature point to determine the scene labels with a greater certainty.

In the early work of relaxation labeling, the labeling is discrete [217]. For a given scene, the relaxation labeling algorithm assigns each component a set of possible interpretations. The first step is to examine constraints between label pairs on connected segments and remove inconsistent labels. The second step is to extend the examination of constraints from local to the whole scene. However, this type of algorithm cannot guarantee that the final labeling is unambiguous, as some objects or components may have more than one consistent label or have no label. Hummel and Zucker developed a probabilistic relaxation to solve this problem [121]. In their algorithm, an object or a component is assigned a set of possible labels with associated weights in the range from 0 to 1. Usually, these weights are set to the conditional probabilities for assignments of the labels. The initial labeling probabilities can be assigned according to the Bayes theorem and based on the assumption that the objects or components in the scene are independent. Then, the labeling probabilities are updated by maximizing similarity or compatibility functions in each iterative loop.

Generally, a relaxation process will find a local maximum on the basis of the particular criterion. However, the relaxation labeling algorithms converge to good results provided that the initial labeling probabilities are reasonable, especially with the Highest-Confidence-First procedure [61, 62]. Li has shown that the solution quality obtained by relaxation labeling is close to that obtained by SA with much less computation [159]. In our experiments, we will use relaxation labeling.

5.3 Recognition of Handwritten Numerals Using MRF Models

As MRF models are multi-dimensional in nature and are able to combine both statistical and structural information, we apply MRF models to recognizing handwritten numerals.

5.3.1 Feature

In Chapters 3 and 4, we introduced methods for recognizing handwritten numerals based on features extracted from outer contours and achieved good results. The descriptors based on contours are also widely used in general 2-D pattern recognition when shapes can be represented by single closed curves. However, some characters cannot be represented by single closed curves, such as i and j; some symbols share the same shapes of outer contours, such as $\odot$, $\oplus$ and $\otimes$. It is necessary to develop a system that is able to recognize other characters and symbols as well as handwritten numerals. We use MRF models with features extracted from skeletons for handwriting recognition.

Thinning is useful in shape representation as well as in data compression. Many thinning algorithms are available in the literature [150]. However, most of them are not robust to noise and cannot deal with holes filled by thick line-segments (shown in Figure 5.10). Such algorithms may introduce distortions such as line-segment displacements at cross regions and filled holes, and false skeletal branches. Recently, Cai and Liu proposed to use 2-D Gabor filters for skeletonization [40]. The parameters of Gabor filters are able to adapt to the width of line-segments. Therefore, Gabor filters can produce features that better serve handwriting recognition tasks. In this system, we will use the thinning algorithm described in Chapter 2. Figure 5.10 gives some results of the skeletonization using Gabor filters.

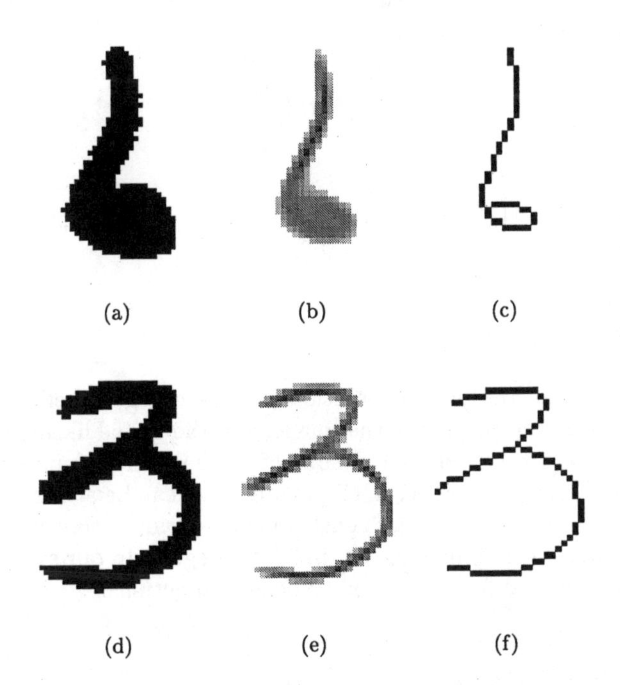

FIGURE 5.10. Skeletonization using Gabor filters. (a) & (d) original images. (b) & (e) orientation energies of Gabor filters. (c) & (f) skeletons.

As we know, many features can be used to describe skeletons. In this system, we used locations and orientations of short segments as features, where a short segment contains five points including two overlapping points with its neighboring segments. The orientation of a segment (D_i) is encoded into one of the eight directions shown in Figure 5.11. However, the cross points have no orientation or have more than one orientation, which cannot be encoded into any of the eight directions. In our experiments, we find that there are two types of cross points: One has three branches and another has four. Therefore, we encode the cross points with three branches into the $8th$ codeword and the cross points with four branches into the $9th$ codeword. Thus, an observation or the feature of a short segment $\mathbf{o}_i$ can be represented by

$$\mathbf{o}_i = \{(x_i, y_i), d_i\}, \tag{5.46}$$

where (x_i, y_i) is the middle point location of the line-segment i and d_i is an index of the codeword for D_i. The index ranges from 0 to 9.

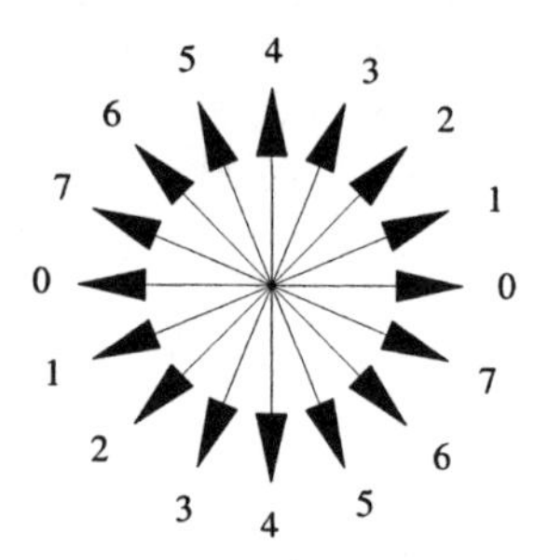

FIGURE 5.11. Orientation codewords for short line segments.

5.3.2 *Clique Function*

As we discussed in the previous sections, clique energy functions are able to describe local information in the image, and the global likelihood energy function can be expressed as the sum of clique energy functions in the neighborhood system. In general, clique functions can be arbitrarily chosen as long as they meet the energy constraint: they must decrease the values of the functions with an increase in matching degree. In our system, clique functions are derived from probability density functions.

A: Statistical Information

The statistical information is modeled by observations of individual states. In order to simplify the computation, we assume that the two components of an observation are independent and the probability density function of

$\{x_{X_i}, y_{X_j}\}$ is of the joint Gaussian form. Thus, we have

$$V_{c1}(\mathbf{o}_i|X_i) = \frac{1}{2}(x_i - \overline{x}_{X_i}, y_i - \overline{y}_{X_i})\Gamma_{2X_i}^{-1}\begin{pmatrix} x_i - \overline{x}_{X_i} \\ y_i - \overline{y}_{X_i} \end{pmatrix} + C_2(X_i) - \log p(d_i|X_i), \tag{5.47}$$

where $C_2(X_i) = \log(2\pi|\Gamma_{2X_i}|^{\frac{1}{2}})$, and Γ_{2X_i} is the covariance matrix.

B: Structural Information

The structural information is modeled by relative spatial positions and relative orientations between states. The relative spatial position between states X_i and X_j is defined as

$$\{x_{ij}, y_{ij}\} = \{x_j - x_i, y_j - y_i\}.$$

The relative orientation between the two states is

$$\Delta D_{ij} = D_j - D_i,$$

which is quantized into one of 16 directions as shown in Figure 5.12, because it is difficult to use Gaussian functions or mixtures of Gaussian functions to model the distributions of orientations [36]. If i or j is a cross point, D_i or D_j will be replaced by the orientation from i to j. For the same reason stated above, we assume that the two components of structural information are independent. Therefore, we have

$$\begin{aligned} V_{c1}(X_i, X_j) &= \mu_{ij}\Big[\frac{1}{2}(x_{ij} - \overline{x}_{X_iX_j}, y_{ij} - \overline{y}_{X_iX_j})\Gamma_{1X_iX_j}^{-1}\begin{pmatrix} x_{ij} - \overline{x}_{X_iX_j} \\ y_{ij} - \overline{y}_{X_iX_j} \end{pmatrix} + \\ & C_1(X_i, X_j) - \log A_{X_iX_j} - \log p(B_{ij}|X_i, X_j)\Big], \end{aligned} \tag{5.48}$$

where μ_{ij} is the weight, $C_1(X_i, X_j) = \log(2\pi|\Gamma_{1X_iX_j}|^{\frac{1}{2}})$, $\Gamma_{1X_iX_j}$ is the covariance matrix, $A_{X_iX_j}$ is the probability that X_j is a neighbor of X_i, and B_{ij} is the index of relative orientation codebooks for ΔD_{ij}.

5.3.3 Maximizing the Global Compatibility

As we stated before, the recognition process in this system is to minimize the global energy function of MRF. We use relaxation labeling for optimization due to its computational efficiency as compared to the simulated annealing algorithm. As the energy function is not suitable to be directly used in relaxation labeling [160, 35], we convert the energy function to the compatibility function. Therefore, maximizing the global compatibility function of MRF is equivalent to minimizing the global energy function.

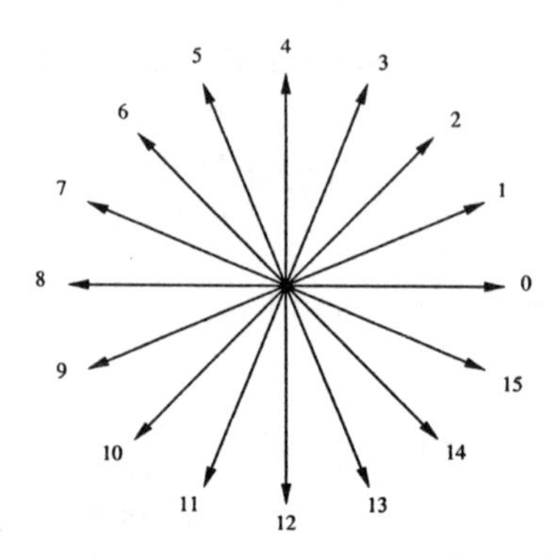

FIGURE 5.12. Relative orientation codebook.

A: Relaxation Labeling Algorithm

Relaxation labeling is an iterative updating algorithm. In the continuous relaxation labeling, the labeling assignment is defined as

$$\mathbf{f} = \{f_{i,X_i} \in [0,1] | \sum_{X_i} f_{i,X_i} = 1, 1 \leq i \leq I\}, \tag{5.49}$$

where f_{i,X_i} is the probability that the best state of short line-segment i is state X_i. According to the theory of relaxation labeling, we can use the following fixed-point iteration scheme [217, 121]:

$$f^{t+1}_{i,X_i|X_i=k} = \frac{f^t_{i,X_i|X_i=k}(1 + q^t_{i,X_i})}{\sum_{X_i} f^t_{i,X_i}(1 + q^t_{i,X_i})}, \tag{5.50}$$

where f^t_{i,X_i} represents the ambiguous relaxation labeling of f_{i,X_i} after t iterations, q^t_{i,X_i} is called support function and is assumed $|q^t_{i,X_i}| \leq 1$. In the case of $q^t_{i,X_i} \geq 0$, the following fixed point iteration scheme can be used

$$f^{t+1}_{i,X_i|X_i=k} = \frac{f^t_{i,X_i|X_i=k} q^t_{i,X_i|X_i=k}}{\sum_{X_i} f^t_{i,X_i} q^t_{i,X_i}}. \tag{5.51}$$

The support function q^t_{i,X_i} can be derived from compatibility functions. The next issue is when the algorithm should terminate. As the final labeling is unambiguous, some heuristic conditions must be set to terminate the algorithm. In some applications, the algorithm terminates if the number of iterations reach a fixed constant or the highest labeling probability for each node exceeds a pre-determined value. In our system, a hybrid method with the highest labeling probability first was used to terminate the relaxation labeling. At any iteration, if the highest state labeling probability for a short line-segment exceeds $1 - \delta$ ($\delta << 1$), the state labeling for that line-segment is set to be unambiguous. If a state labeling probability for a short line-segment is smaller than δ, the probability of labeling the line-segment to that state is set to zero. If the fixed number of iterations is reached, the relaxation labeling terminates and the winner-take-all strategy is used. The winner-take-all strategy can be described as that if $f_{i,X_i} \in (0,1)$ is the highest among ambiguous labels, we set

$$\begin{aligned} f_{i,X_i|X_i=j} &= 1, \\ f_{i,X_i|X_i=k} &= 0 \qquad k \neq j. \end{aligned} \tag{5.52}$$

B: Maximizing the Global Compatibility

The support function in (5.50) and (5.51) must be a monotonically increasing function of matching degree. According to the energy constraint of clique functions, however, the support function is a decreasing function of energy. As a result, the energy functions cannot be used directly in relaxation labeling. Therefore, we convert the energy functions into compatibility functions defined as follows:

$$\begin{aligned} H_{c1}(X_i, X_j) = \mu_{ij} Con - V_{c1}(X_i, X_j); \\ H_{c2}(\mathbf{o}_i|X_i) = Con - V_{c2}(\mathbf{o}_i|X_i), \end{aligned} \tag{5.53}$$

where Con is a constant, and $\sum_{j \in \mathcal{N}_i} \mu_{ij} = 1$. The global compatibility function is:

$$G(\mathbf{O}, \mathbf{X}, \lambda_w) = \sum_i \sum_{X_j} \left[H_{c2}(\mathbf{o}_i|X_i) f_{i,X_i} + \sum_{j \in \mathcal{N}_i} \sum_{X_j} H_{c1}(X_i, X_j) f_{i,X_i} f_{j,X_j} \right]. \tag{5.54}$$

The relationship between $U(\mathbf{O}, \mathbf{X}, \lambda_w)$ and $G(\mathbf{O}, \mathbf{X}, \lambda_w)$ can be derived from (5.53) and (5.54)

$$U(\mathbf{O}, \mathbf{X}, \lambda_w) = 2I \cdot Con - G(\mathbf{O}, \mathbf{X}, \lambda_w). \tag{5.55}$$

Clearly, minimizing the global energy function is exactly the same as maximizing the global compatibility function. Therefore, the support function q^t_{i,X_i} can be obtained from the gradient information given by

$$\begin{aligned} q^t_{i,X_i} &= \frac{\partial G^t(\mathbf{O}, \mathbf{X}, \lambda_w)}{\partial f^t_{i,X_i}} \\ &= H_{c2}(\mathbf{o}_i|X_i) + \sum_{j \in \mathcal{N}_i} \sum_{X_j} H_{c1}(X_i, X_j) f_{j,X_j}. \end{aligned} \tag{5.56}$$

In this way, we can obtain the best state set and the minimum of the global energy function $U(\mathbf{O}, \mathbf{X}_{best}, \lambda_w)$. The recognition decision can be made based on the global optimal energy.

5.3.4 Experimental Results

The performance of the proposed approach is evaluated on the standard subset of CEDAR CDROM1 database used in Chapters 3 and 4. As the images in the "goodbs" set are well segmented, the performance on this test set will demonstrate the effectiveness of the system described in this chapter. We conducted our experiments on the "goodbs" set. All images after pre-processing were normalized to fit in a 40×50 pixel box. In our experiments, a recognition rate of 96.48% was achieved using only one model per pattern class without rejection. This performance is comparable to the best results published recently [154, 151, 20]. Table.5.1 gives the best results obtained by neural networks for handwritten digit recognition. Figure 5.13 shows the comparison of performance among different approaches, where 'FD' denotes the method based on Fourier descriptors [170], 'FFD' the fuzzy feature description [170], 'MCNN' the multi-layer cluster neural network [154], 'RBNN' the radial basis function neural network [120], and 'MRF' is for MRFs. In these methods, some use several models per class or combine several classifiers to recognize handwritten numerals. For instance, Hwang and Bang used 50 models per class to achieve 97.90% recognition rate [120] . It is difficult to directly compare the results of different methods on different databases; however, the performance of the MRF-based method is very high with only one model per class in our experiments. If using multiple models per class, it is expected that the performance of the proposed method can be further improved.

TABLE 5.1. The best results of networks [20]

Method	Error rate
PNN	2.5
MLP	4.3
RBF	3.9

It is well known that the MRF is a computation-intensive model in low-level image processing. In our system, however, the MRF models were performed on feature segments instead of raw image pixels. Because there are only 15×45 feature segments in a 40×50 size-normalized image, the system can recognize two digits per second on average using an old Silicon Graphics computer (4×33MHz, IP7 processors).

5.4 Conclusion

In this Chapter, we have introduced the concept of states into MRF models and developed a new approach based on MRF models for pattern and handwriting recognition. We applied this approach to the recognition of un-

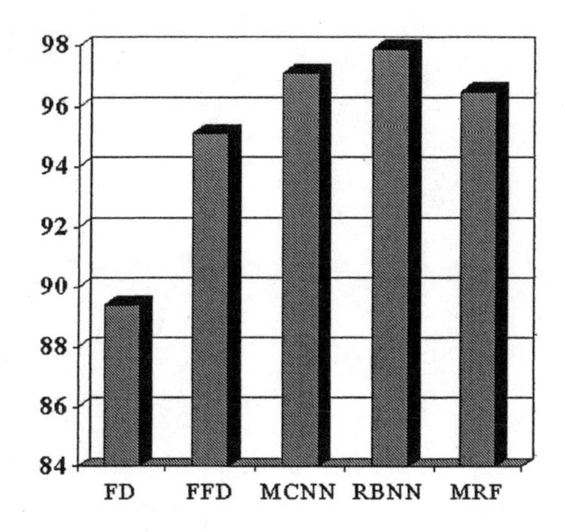

FIGURE 5.13. The comparison of performance.

constrained handwritten numerals. The features used in the experiments were extracted from image skeletons by using Gabor filters. The experimental results show the advantages of the MRF-based approach in handwriting recognition.

In the MRF framework, we can model the statistical information by individual states and the structural information by the relationships between states. Therefore, this approach combines both statistical and structural information. Conventional statistical approaches cannot model structural information and conventional structural approaches are not able to capture the statistical information. Moreover, the MRF-based approach does not need to extract *grammars* for structural representations, which is a difficult task. It is also able to deal with structural variations, because the structural information is modeled by the Bayesian paradigm.

We would also like to point out that it is not difficult to see that HMMs are a special case of MRF models used in this chapter. The neighborhood system of the HMM is same as that of the MRF model when two boundary neighborhoods are considered in one-dimensional space. Also, the state transition in HMMs is a special case of the state relationship in MRF models.

6

Markov Random Field Models for Recognizing Handwritten Words

This chapter presents a system for recognizing unconstrained handwritten words. In this system, we describe handwritten words in terms of directional line segments that are extracted by Gabor filters and we use MRF models for recognizing unconstrained handwritten words. As discussed in detail in Chapter 5, the main advantage of MRF models is that they provide a flexible and natural framework for modeling the interaction between spatially related random variables in their neighborhood systems. To deal with shape variations in handwritten words, we use fuzzy neighborhood systems and fuzzy matching measurements. The relaxation labeling algorithm is used to maximize the global compatibilities of the MRF models. We investigate the influence of neighborhood size and iteration number of relaxation labeling on recognition rates.

In this system, the features are the parameters of word image line segments. A line segment has endpoints at pixels $P1(x_1, y_1)$ and $P2(x_2, y_2)$ and is characterized by its orientation θ, length l, and line centroid $c(x_0, y_0)$:

$$\theta = \arctan(\frac{y_2 - y_1}{x_2 - x_1}),$$

$$l = \sqrt{(x_2 - x_1)^2 + (y_2 - y_1)^2},$$

$$x_0 = \frac{(x_2 + x_1)}{2}, \quad y_0 = \frac{(y_2 + y_1)}{2}.$$

We use the Gabor filter to extract line segments, whose parameters are determined by the method described in Chapter 2. First, using the Gabor filter we calculate the power at each point from the complex response of a word image, which generates a power image at an orientation specified by the Gabor filter. Then we extract oriented line segments from the thresholded power image. Figure 2.30 shows examples of extracted line segments.

6.1 Markov Random Field for Handwritten Word Recognition

Off-line handwritten word recognition is a very difficult problem as the dynamic information, such as the order of strokes, is not readily available.

In addition to noise, deformation and different writing styles further hinder off-line handwriting recognition. In the following sections, we will present a MRF-based approach to recognizing handwritten words.

6.1.1 Markov Random Field for Structural Modeling

Consider a structure $\mathbf{S}$ of a handwritten word, where

$$\mathbf{S} = \{1, 2, \cdots, i, \cdots, m\} \tag{6.1}$$

consists of a set of line-segments denoted by digits, where m is the number of line-segments in structure $\mathbf{S}$. Let $\mathbf{S}^+$ denote structure $\mathbf{S}$ plus a NULL line-segment denoted by 0. Define a neighborhood system on $\mathbf{S}$

$$\mathcal{N} = \{\mathcal{N}_i | \forall i \in \mathbf{S}\} \tag{6.2}$$

where $\mathcal{N}_i$ is the collection of line-segments that are the neighbors of the line-segment denoted by i. A clique c defined on $(\mathbf{S}, \mathcal{N})$ is a region or a singleton.

Let $\mathbf{R}$ be a set of structures of given templates, where

$$\mathbf{R} = \{\mathbf{R}_1^+, \mathbf{R}_2^+, \cdots, \mathbf{R}_w^+, \cdots, \mathbf{R}_W^+\} \tag{6.3}$$

are W sets of labels and W is the number of all templates. For a given structure $\mathbf{R}_w$,

$$\mathbf{R}_w = \{1, 2, \cdots, j, \cdots, M_w\} \tag{6.4}$$

consists of a set of labels denoted by digits that symbolize the lines in the structure of a given template and $\mathbf{R}_w^+$ is the structure $\mathbf{R}_w$ plus a NULL label denoted by 0, where M_w is the number of line-segments in the template $\mathbf{R}_w$. As the structures of handwritten words are usually deformed, this will result in some unmatched line-segments. Therefore, the NULL label is necessary to cover unmatched line-segments.

During the recognition stage, the lines in $\mathbf{S}$ are assigned a set of labels $\mathbf{F}$:

$$\mathbf{F} = \{F_1, F_2, \cdots, F_m\} \tag{6.5}$$

which is a family of random variables defined on $\mathbf{S}$. $\mathbf{F}$ is an MRF on $\mathbf{S}$ with respect to $\mathcal{N}$ if and only if

- $P(\mathbf{F} = \mathbf{f}) > 0, \qquad \forall \mathbf{f};$
- $P(f_i | f_j, i \neq j \ \&\ \forall j \in \mathbf{S}) = P(f_i | f_j, j \in \mathcal{N}_i),$

where $P(\cdot)$ and $P(\cdot|\cdot)$ are the joint and conditional probability distribution functions (pdf's), respectively. In this chapter, the joint pdf is assumed to be of the form:

$$P(\mathbf{F} = \mathbf{f}) = Z^{-1} \times e^{-U(\mathbf{f})/T}, \tag{6.6}$$

and

$$U(\mathbf{f}) = \sum_{c \in \mathbf{C}(\mathbf{S}, \mathcal{N})} V_c(\mathbf{f}), \tag{6.7}$$

where Z is the partition function, T is the temperature, $U(\mathbf{f})$ is the energy function, and $V_c(\mathbf{f})$'s are the clique functions. The MRF model provides a unique framework which connects the global optimum with local properties via clique functions designed on the neighborhood system. Labeling algorithms, which are used to minimize the energy function $U(\mathbf{f})$, can derive the global optimum from local properties.

According to the theory of structural representation, word shapes can be represented by two types of features: *fragmental* features such as the location and orientation of a line-segment; *configurational* features that are based on the relationships among the fragmental features. Studies have shown that during reading, the first stage is to acquire word shapes and features necessary for decoding the word [241]. The human visual system uses mostly configurational features rather than fragmental features to recognize letters and words [241]. MRFs is able to model both types of features that represent the structure of the word: the first-order clique functions for the fragmental features and the higher-order clique functions, which describe the spatial relationships among the line-segments, for the configurational features. Due to variations in writing styles, writing tools and writing conditions, there are considerable uncertainties in structures of handwritten words, which are not statistical by nature. In order to deal with such uncertainties, we define MRF models on fuzzy neighborhoods and design clique functions based on fuzzy measurements.

6.1.2 Recognition based on Maximum a posteriori *Probability*

In general, handwritten word recognition is based on the maximum *a posteriori* (MAP) paradigm, which is a process to assign line-segments optimally in the word structure to the labels in the best matching template. Formally, we have

$$\mathbf{f}^* = \arg_{\mathbf{f}} \max_{w} \max_{\mathbf{f} \in \mathbf{R}_w^+ \in \mathbf{R}} P(\mathbf{F} = \mathbf{f} | \mathbf{R}_w^+, \mathbf{S}), \tag{6.8}$$

$$w^* = \arg_{w} \max_{w} \max_{\mathbf{f} \in \mathbf{R}_w^+ \in \mathbf{R}} P(\mathbf{F} = \mathbf{f} | \mathbf{R}_w^+, \mathbf{S}), \tag{6.9}$$

where $P(\mathbf{F} = \mathbf{f} | \mathbf{R}_w^+, \mathbf{S})$ is the *a posteriori* function for the given template $\mathbf{R}_w^+$ and the word structure. $P(\mathbf{F} = \mathbf{f} | \mathbf{R}_w^+, \mathbf{S})$ is assumed to be the Gibbsian form [179]:

$$P(\mathbf{F} = \mathbf{f} | \mathbf{R}_w^+, \mathbf{S}) = Z^{-1} \times e^{-U(\mathbf{f}; \mathbf{R}_w^+, \mathbf{S})}, \tag{6.10}$$

and

$$U(\mathbf{f}; \mathbf{R}_w^+, \mathbf{S}) = \sum_{c \in \mathbf{C}(\mathbf{S}, \mathcal{N})} V_c(\mathbf{f}; \mathbf{R}_w^+, \mathbf{S}), \tag{6.11}$$

where $T = 1$, $U(\mathbf{f}; \mathbf{R}_w^+, \mathbf{S})$ is the *a posteriori* energy function and $V_c(\mathbf{f}; \mathbf{R}_w^+, \mathbf{S})$'s are the *a posteriori* clique functions.

6.2 Neighborhood Systems and Cliques

In handwritten words, the writing style of an individual character is influenced by its neighboring characters depending on writing styles and the position and motion of the hand during the writing process. Some characters may be strongly influenced by their neighboring characters, but almost not influenced by non-neighboring characters. For instance, the character "i" in Figure 6.1(a) are strongly influenced by its neighboring characters "o" and "n". While the character "o" in Figure 6.1(b) almost has no influence on "i" in the same word. There are two types of the interfering relationships between line-segments:

Inter-character relationship: the interfering relationships between line-segments of neighboring characters are stronger than those between line-segments of non-neighboring characters;

Intra-character relationship: due to intrinsic shapes of individual characters, the correlative relationships between intra-character line-segments are stronger than those between inter-character line-segments.

(a) (b)

FIGURE 6.1. Examples of characters influenced by their neighboring characters.

In order to deal with uncertainties in the strength of interfering relationships, we use a fuzzy neighborhood system. We define the membership of the neighborhood in x-axis as

$$FN_x(x_i, x_j) = \begin{cases} 1 & |x_i - x_j| \leq \sigma_x, \\ 2 - \frac{|x_i - x_j|}{\sigma_x} & \sigma_x \leq |x_i - x_j| < 2\sigma_x, \\ 0 & |x_i - x_j| \geq 2\sigma_x, \end{cases} \tag{6.12}$$

where x_i is x-axis coordinate of line-segment i in $\mathbf{S}$, x_j is x-axis coordinate of line-segment j in $\mathcal{N}_i$ and σ_x is a constant. According to the intra-character relationship, the relationships between intra-character line-segments are strong. Therefore, we can set $FN_x(x_i, x_j) = 1$, where i and j are the line-segments belonging to a character. As the line-segments of a character locate within the width of the character, we have

$$\sigma_x \approx W_c \tag{6.13}$$

where W_c is the average width of individual characters in the word.

As the number of characters in the word is usually unknown, W_c is approximated by $W_c = W_S / C_w$, where W_S is the width of the testing word and C_w is the number of characters in the template. Similarly, the range of neighborhood in y-axis is defined as

$$FN_y(y_i, y_j) = \begin{cases} 1 & |y_i - y_j| \leq \sigma_y, \\ 2 - \frac{|y_i - y_j|}{\sigma_y} & \sigma_y \leq |y_i - y_j| < 2\sigma_y, \\ 0 & |y_i - y_j| \geq 2\sigma_y, \end{cases} \tag{6.14}$$

where y_i is y-axis coordinate of line-segment i in $\mathbf{S}$, y_j is y-axis coordinate of line-segment j in $\mathcal{N}_i$ and σ_y is a constant. If $|x_i - x_j|$ is small, the line-segments along the y-axis direction may belong to one character, $FN_y(y_i, y_j) \approx 1$. Therefore, it is appropriate to set σ_y to the word height.

The membership of neighborhood related to the line-segment (x_i, y_i) shown in Figure 6.2 is defined as

$$FN(i, j) = FN_x(x_i, x_j) FN_y(y_i, y_j). \tag{6.15}$$

One of the major advantages of MRF is its flexibility in designing cliques that can be in terms of singleton (first order), second order, or higher-order neighborhood systems. For two dimensional irregular sites, the structure can be represented by the spatial relationships between nodes, where a node denotes a line-segment and the location of the node is the centroid of the line-segment. For example, the spatial relationship between nodes b and c in Figure 6.3 can be derived from that between nodes a and b and between nodes a and c. Therefore, the global structure in 2-D irregular sites can be modeled by MRFs based on second-order cliques. In this chapter, we consider only second-order (two-node) cliques, because they are able to effectively and efficiently represent the interaction between random variables for handwritten-word recognition.

6.3 Clique Functions

We may choose clique functions arbitrarily as long as they meet the energy constraint: If the features in a clique match the features in the given template, the energy function decreases; otherwise, the energy function increases. According to the energy constraint, we define the clique function

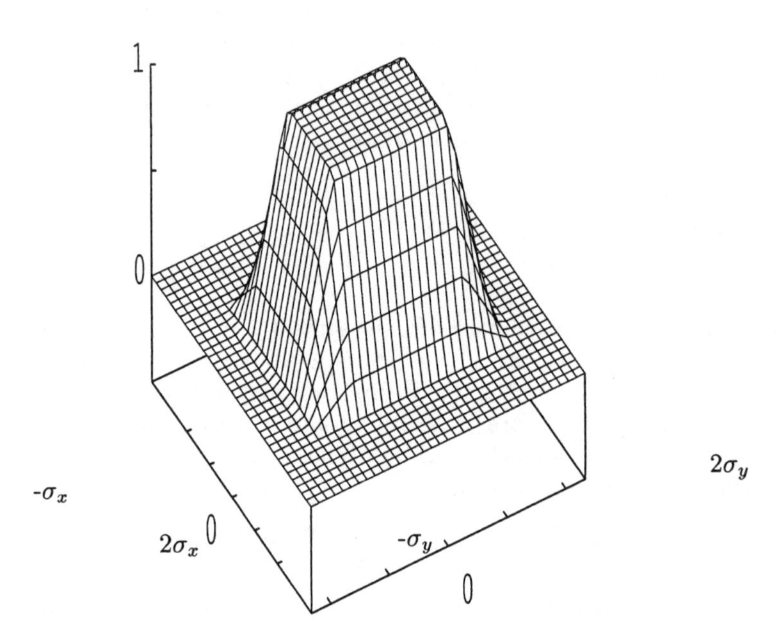

FIGURE 6.2. Fuzzy membership of neighborhood related to a line-segment.

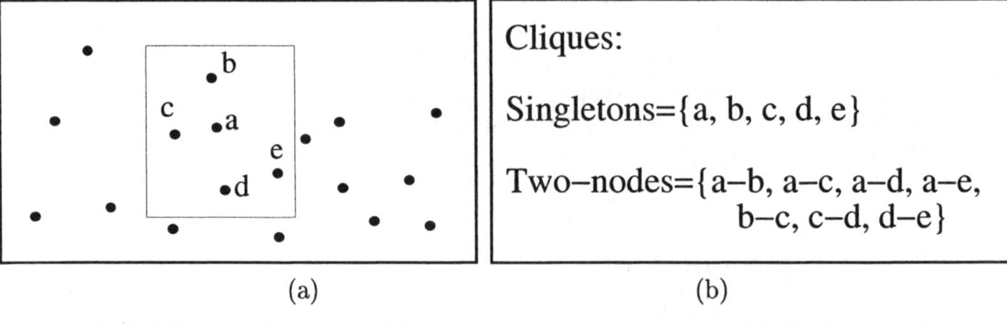

FIGURE 6.3. (a) Neighborhood system on irregular sites; (b) Singleton and two-node cliques on irregular sites.

associated with line-segment i as follows:

$$V_c(u,v|i,j;\mathbf{S},\mathbf{R}_w^+) = [1-\mu(u,v|i,j;\mathbf{S},\mathbf{R}_w^+)B_1(u|i;\mathbf{S},\mathbf{R}_w^+)B_2(u|i;\mathbf{S},\mathbf{R}_w^+)] \\ B_3(i,j;\mathbf{R}_w^+), \tag{6.16}$$

where $j \in \mathcal{N}_i$, $u = f_i$, $v = f_j$, $\mu(\cdot)$ is a fuzzy matching measurement between the features in a clique and the features in the template $\mathbf{R}_w^+$; and $B_1(\cdot)$, $B_2(\cdot)$ and $B_3(\cdot)$ are non-negative weights. For convenience, we denote $\mu(u,v|i,j;\mathbf{S},\mathbf{R}_w^+)$, $B_1(u|i;\mathbf{S},\mathbf{R}_w^+)$, $B_2(u|i;\mathbf{S},\mathbf{R}_w^+)$ and $B_3(i,j;\mathbf{S},\mathbf{R}_w^+)$ as $\mu(u,v|i,j)$, $B_1(u|i)$, $B_2(u|i)$ and $B_3(i,j)$, respectively. $\mu(u,v|i,j)$ is defined as

$$\mu(u,v|i,j) = \frac{\mu_l(u,v|i,j)+\mu_d(u,v|i,j)+\mu_a(u,v|i,j)}{3} \tag{6.17}$$

where $\mu_l(u,v|i,j)$, $\mu_d(u,v|i,j)$ and $\mu_a(u,v|i,j)$ are fuzzy matching measurements of length, distance and angle, respectively, which have the following forms:

$$\mu_l(u,v|i,j) = \begin{cases} 1 & C_l(l_{Si},l_{Sj},l_{Ru},l_{Rv}) \geq T_h, \\ C_l(l_{Si},l_{Sj},l_{Ru},l_{Rv})/T_h & \text{else,} \end{cases} \tag{6.18}$$

$$\mu_d(u,v|i,j) = \begin{cases} 1 & C_d(l_{Si},l_{Sj},l_{Ru},l_{Rv}) \geq T_h, \\ 1.25C_d(l_{Si},l_{Sj},l_{Ru},l_{Rv})/T_h & \text{else,} \end{cases} \tag{6.19}$$

$$\mu_a(u,v|i,j) = \begin{cases} 0 & \Delta A(u,v|i,j) \geq \frac{\pi}{2}, \\ \cos[\Delta A(u,v|i,j)] & \text{else,} \end{cases} \tag{6.20}$$

where T_h is a threshold, l_{Si} and l_{Sj} are lengths of line-segment i and j in $\mathbf{S}$, l_{Ru} and l_{Rv} are lengths of line-segment u and v in $\mathbf{R}_w^+$, and $C_l(l_{Si},l_{Sj},l_{Ru},l_{Rv})$, $C_d(l_{Si},l_{Sj},l_{Ru},l_{Rv})$, and $\Delta A(u,v|i,j)$ are

$$C_l(l_{Si},l_{Sj},l_{Ru},l_{Rv}) = \frac{\min\{l_{Si}/l_{Sj}, l_{Ru}/l_{Rv}\}}{\max\{l_{Si}/l_{Sj}, l_{Ru}/l_{Rv}\}}, \tag{6.21}$$

$$C_d(l_{Si},l_{Sj},l_{Ru},l_{Rv}) = \frac{\min\{\log d_{ij}, \log d_{uv}\}}{\max\{\log d_{ij}, \log d_{uv}\}}, \tag{6.22}$$

$$\Delta A(u,v|i,j) = |\angle_{ij} - \angle_{uv}|, \tag{6.23}$$

where d_{ij} is the distance between segment i and j in $\mathbf{S}$, d_{uv} is the distance between segment u and v in $\mathbf{R}_w^+$, $\angle_{ij}$ is the angle between segment i and j in $\mathbf{S}$ and $\angle_{uv}$ is the angle between segment u and v in $\mathbf{R}_w^+$. If u or v (or both) is NULL, the matching can not be measured. A simple remedy is to set μ to a small constant; we set $\mu = 0.3$. $B_1(u|i)$ and $B_2(u|i)$ will be discussed later. $B_3(i,j)$ is a normalized weight that determines the importance of the given clique function:

$$\sum_{c \in \mathbf{C}(\mathbf{S},\mathcal{N})} B_3(i,j) = 1. \tag{6.24}$$

Because $\mu(u,v|i,j)$ represents the degree of matching between the features of the clique and the template, the clique function defined in (6.16) satisfies the energy constraint.

6.4 Maximizing the Compatibility with Relaxation Labeling

The recognition process of the proposed method is to minimize the energy of MRF using relaxation labeling. However, as discussed in Chapter 5, the energy function is not suitable to be directly used in relaxation labeling. For this, we convert the energy function to a compatibility function. We show that maximizing the compatibility of MRF is equivalent to minimizing the energy of MRF.

6.4.1 Relaxation Labeling

Relaxation labeling has been successfully applied to boundary detection [106], scene labeling [217] and handwritten numeral recognition. The theory and iterative updating algorithms were developed by Kittler [144] and Hummel and Zucker [121]. In this chapter, we use the ambiguous relaxation labeling algorithm proposed by Hummel and Zucker, as it generally leads to the global optimal point [121].

In the ambiguous relaxation labeling, we use the labeling assignment to denote an MRF labeling configuration [160]. The labeling assignment is defined as

$$\mathbf{f} = \{f_{i,u} \in [0,1] | \sum_{u \in \mathbf{R}_w^+} f_{i,u} = 1, i \in \mathbf{S}, u \in \mathbf{R}_w^+\} \tag{6.25}$$

where $f_{i,u}$ is the probability of assigning a line-segment i in $\mathbf{S}$ to the label u in $\mathbf{R}_w^+$. According to the theory of relaxation labeling, we use the following fixed point iteration scheme [121, 217]:

$$f_{i,u}^{t+1} = \frac{f_{i,u}^t(1+q_{i,u}^t)}{\sum_{k \in \mathbf{R}_w^+} f_{i,k}^t(1+q_{i,k}^t)} \tag{6.26}$$

where $f_{i,u}^t$ represents the ambiguous relaxation labeling of $f_{i,u}$ after t times iterations, $q_{i,u}^t$ is called support function and is assumed $|q_{i,u}^t| \leq 1$. If $q_{i,u}^t \geq$

0, the following fixed point iteration scheme can be used

$$f_{i,u}^{t+1} = \frac{f_{i,u}^t q_{i,u}^t}{\sum\limits_{k \in \mathbf{R}_W^+} f_{i,k}^t q_{i,k}^t}. \tag{6.27}$$

Usually, the number of iterations for relaxation labeling is limited, and the final labeling must be unambiguous. Therefore, the winner-take-all strategy is used to obtain a set of unambiguous labels. After several iterations, if $f_{i,u} \in (0,1)$ is the highest among ambiguous labels, we set

$$\begin{aligned} f_{i,u} &= 1, & \\ f_{i,v} &= 0 & v \neq u, \\ f_{j,u} &= 0 & u \neq 0 \ \&\ j \neq i. \end{aligned} \tag{6.28}$$

6.4.2 Maximizing the Compatibilities

During the process of relaxation labeling, if the features of the line-segment i and its neighborhood match the line-segment u and its neighborhood in the template $f_{i,u}$ should increase. Therefore, the support function in (6.26) and (6.27) must be a monotonically increasing function of matching degree. While, according to the energy constraint of clique functions, the support function is a decreasing function of the *posterior* energy. As a result, it is inconvenient to use the energy function for relaxation labeling. We convert the energy function into the compatibility function which represents the matching degree between the handwritten word and the given template. The global compatibility function is defined as follows:

$$G(\mathbf{f}; \mathbf{R}_w^+, \mathbf{S}) = 1 - U(\mathbf{f}; \mathbf{R}_w^+, \mathbf{S}). \tag{6.29}$$

From (6.16) and (6.24), we obtain

$$G(\mathbf{f}; \mathbf{R}_w^+, \mathbf{S}) = \sum_{c \in \mathbf{C}(\mathbf{S}, \mathcal{N})} H_c(\mathbf{f}; \mathbf{R}_w^+, \mathbf{S}), \tag{6.30}$$

where $H_c(\mathbf{f}; \mathbf{R}_w^+, \mathbf{S})$ is called clique compatibility function, and is derived from (6.16) and (6.24)

$$H_c(u, v|i, j; \mathbf{S}, \mathbf{R}_w^+) = \mu(u, v|i, j) B_1(u|i) B_2(u|i) B_3(i, j). \tag{6.31}$$

For ambiguous relaxation labeling, we can rewrite (6.30) as follows:

$$G(\mathbf{f}; \mathbf{R}_w^+, \mathbf{S}) = \sum_{i \in \mathbf{S}} \sum_{j \in \mathbf{S}} \sum_{u \in \mathbf{R}_w^+} \sum_{v \in \mathbf{R}_w^+} H_c(u, v|i, j; \mathbf{S}, \mathbf{R}_w^+) f_{i,u} f_{j,v}. \tag{6.32}$$

Now, the problem of handwritten-word recognition becomes that of maximizing the compatibilities of MRFs, and (6.8) and (6.9) become

$$\mathbf{f}^* = \arg_{\mathbf{f}} \max_{w} \max_{\mathbf{f} \in \mathbf{R}_w^+ \in \mathbf{R}} G(\mathbf{f}; \mathbf{R}_w^+, \mathbf{S}), \tag{6.33}$$

$$w^* = \arg_w \max_w \max_{\mathbf{f} \in \mathbf{R}_w^+ \in \mathbf{R}} G(\mathbf{f}; \mathbf{R}_w^+, \mathbf{S}). \tag{6.34}$$

If the relaxation labeling for maximizing the compatibility of MRF is based on the gradient information $\nabla G(\mathbf{f})$, the support function in (6.26) and (6.27) is given by

$$\begin{aligned} q_{i,u}^t &= \frac{\partial G^t(\mathbf{f}; \mathbf{R}_w^+, \mathbf{S})}{\partial f_{i,u}^t} \\ &= \sum_{j \in \mathbf{S}} \sum_{v \in \mathbf{R}_w^+} [H_c(u, v|i, j; \mathbf{S}, \mathbf{R}_w^+) + H_c(v, u|j, i; \mathbf{S}, \mathbf{R}_w^+)] f_{j,v}^t. \end{aligned} \tag{6.35}$$

In this way, handwritten-word recognition can be performed by maximizing the compatibilities of MRFs.

6.5 Design of Weights

In low-level image processing, most MRF models are based on rectangular lattices and features are extracted from pixels. In handwritten-word recognition applications, we extract features from line-segments. Each segment contains a number of pixels. It will make different contributions to the clique compatibility functions and the global compatibility of MRF. Further, the neighborhood memberships of line-segments also play an important role in determining its contributions. Taking these constraints and(6.24) into consideration, we define $B_3(i, j)$ as follows:

$$B_3(i, j) = \frac{l_{Si}(l_{Si} + l_{Sj}) FN(i, j)}{\sum_{r \in \mathbf{S}} l_{Sr} \sum_{k \in \mathcal{N}_i} (l_{Si} + l_{Sk}) FN(i, j)}. \tag{6.36}$$

Normally, some distortions in features are inevitable. As short line-segments are relatively easier to be distorted by noises, the weight $B_3(i, j)$ is designed to be in proportion to line-segment lengths to maintain a robust performance. Therefore, the global compatibility of MRF is mainly based on more reliable line-segments.

As far as $B_1(u|i)$ is concerned, it is related to the positions of line-segments. If line-segment i in $\mathbf{S}$ is of similar orientation to that of line-segment u in $\mathbf{R}_w^+$, their relationships are defined as follows:

$$\mu_x(u|i) = \begin{cases} 1 & |\frac{x_i}{W_S} - \frac{x_u}{W_R}| \le 0.8 \frac{W_c}{W_R}, \\ 1 - \dfrac{|\frac{x_i}{W_S} - \frac{x_u}{W_R}| - 0.8 \frac{W_c}{W_R}}{1.2 \frac{W_c}{W_R}} & 0.8 \frac{W_c}{W_R} < |\frac{x_i}{W_S} - \frac{x_u}{W_R}| < 2 \frac{W_c}{W_R}, \\ 0 & \text{else}, \end{cases} \tag{6.37}$$

$$\mu_y(u|i) = \begin{cases} 1 & |y_i - y_u| \leq 20, \\ 1 - \dfrac{|y_i - y_u| - 20}{40} & 20 < |y_i - y_u| < 60, \\ 0 & \text{else,} \end{cases} \tag{6.38}$$

where W_S is the width of the given word and W_R is the width of the given template. $B_1(u|i)$ is given by

$$B_1(u|i) = \begin{cases} 0.5[\mu_x(u|i) + \mu_y(u|i)] & u \neq 0 \ \& \ \mu_x(u|i)\mu_y(u|i) > 0, \\ 0.3 & u = 0, \\ 0 & \mu_x(u|i)\mu_y(u|i) = 0. \end{cases} \tag{6.39}$$

Because $B_1(u|i)$ deals with the spatial relationship between line-segment i and label u, we may consider only the possible matching pairs related to i in a certain range where $B_1(u|i) > 0$. As the computation is closely related to the search space and this range is much smaller compared with the entire search space, using $B_1(u|i)$ can significantly reduce the computation.

In the experiments, all line-segments are divided into eight groups with similar orientations. $B_2(u|i)$ provides the interaction between feature groups. Similar to $B_3(i,j)$, the contribution of a line-segment to $B_2(u|i)$ is in proportion to its length, thus we define $B_2(u|i)$ as

$$B_2(u|i) = \frac{\sum_{j \in \mathcal{N}_i} l_{Sj} \max_v \{\mu(u,v|i,j)\}}{\sum_{j \in \mathcal{N}_i} l_{Sj}}, \tag{6.40}$$

where i and j belong to different groups.

6.6 Experimental Results

This section presents the results of handwritten-word recognition using MRFs with the ambiguous relaxation labeling technique. In the following, we will discuss the influence of neighborhood size and iteration number on recognition rate.

6.6.1 Neighborhood Size

The size of neighborhood for a line-segment influences the performance of the handwritten word recognizer. There exists an optimal size of neighborhood under the constraints of inter-character and intra-character relationships. Furthermore, since the computation is closely related to the size of the neighborhood. Therefore, we may choose the neighborhood size as small as possible so long as the recognition rate does not drop.

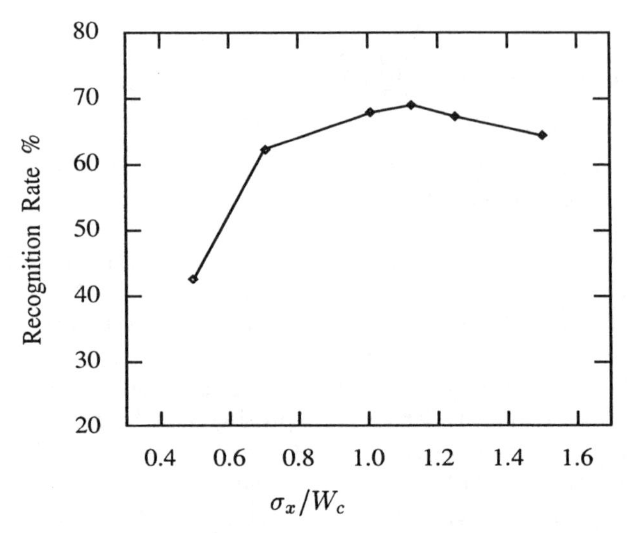

FIGURE 6.4. The relationship between the recognition rates and σ_x.

From Figure 6.4, we find that, if σ_x is too large or too small, the recognition rate of the system will drop, because when σ_x is too small, very few line-segments will be contained in the neighborhood of segment $i \in \mathbf{S}$ and there will not be enough interaction information to properly represent the structure of the word. On other hand, if $\sigma_x$is too large, there are many line-segments having weak relationships with the line-segment i in $\mathcal{N}_i$. A good neighborhood of segment should contain the line-segments belonging to the same character of the segment and having strong relationships with the segment. Figure 6.4 shows that the system obtains the best performance at $\sigma_x/W_c = 1.125$, where the number of iterations is eight and $\sigma_y = 50$.

Figure 6.5 shows the relationship between σ_y and the recognition rate, where the number of iterations is eight and $\sigma_x = 1.125W_c$. The recognition rate increases with σ_y until $\sigma_y = 50$, then it remains unchanged. In fact, when $\sigma_y = 50$, $FN_y(\cdot,\cdot) \approx 1$. This result is consistent with the intra-character relationship. In Figure 6.5, $\sigma_y = \infty$ means $FN_y(i,j) = 1$ for all i and j.

6.6.2 Iterations

In theory, a larger iteration number in the labeling process will result in better recognition performance. However, this will also result in increased computation. Therefore, it is necessary to reduce the search space in order to reduce computation. In our experiments, we divide all segments into eight groups, each of which contains a few line-segments. The MRF with relaxation labeling is performed for each group independently ($B_2(\cdot)$ makes use of the information of the interaction between groups). $B_1(\cdot)$ imposes the position constraints on the search space. Therefore, only a few iterations

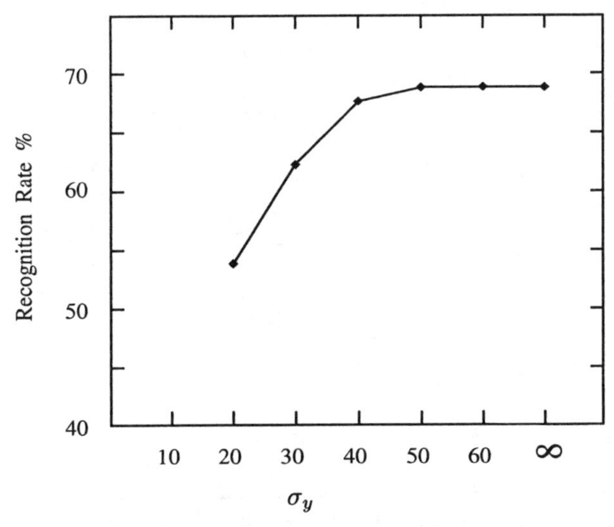

FIGURE 6.5. The relationship between the recognition rates and σ_y.

are needed to achieve a good recognition rate. Figure 6.6 shows that the recognition rate increases rapidly initially and becomes steady after only eight iterations. Table 6.1 gives the handwritten word recognition rates within top five proposals after eight iterations with $\sigma_x/W_c = 1.125$ and $\sigma_y = 50$.

TABLE 6.1. Word recognition accuracies of MRF method for testing

113 testing word images		
first proposal	78	69.0%
among top two proposals	95	84.1%
among top three proposals	102	90.3%
among top four proposals	106	93.8%
among top five proposals	109	96.5%

6.7 Conclusion

In this chapter, we have presented an approach based on MRF models with the ambiguous relaxation labeling technique to recognizing handwritten words. The MRF provides a natural and flexible framework for representing two-dimensional structures. We have designed new clique functions based on fuzzy matching measurements to deal with structural variations. Using the clique functions we can easily convert the problem of maximizing *a posteriori* probability or minimizing *a posteriori* energy to that of maximizing the compatibility of MRF which can be implemented readily

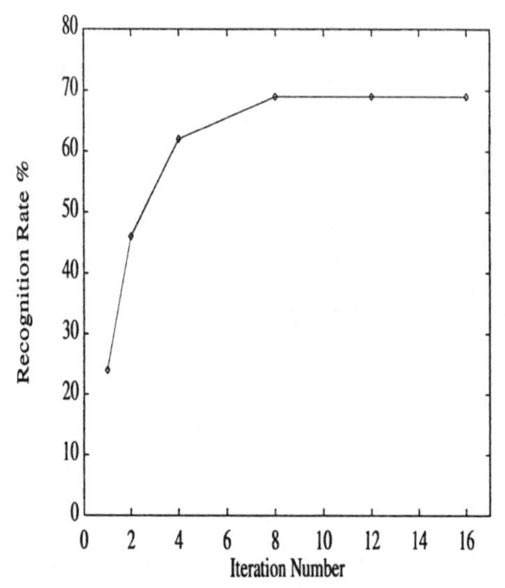

FIGURE 6.6. The relationship between the recognition rates and iteration number.

in the relaxation labeling process. The MRF is a natural framework for the global compatibility via clique compatibility functions defined on the neighborhood system. Therefore, MRF models with relaxation labeling is an effective technique for handwritten-word recognition.

The experimental results were obtained on average using 7.5 training images per word. We were able to achieve recognition rates from 69.0% to 96.5% among top 1 to top 5 positions, which are comparable to the recently published results for which the researchers used much larger training sets. For instance, the recognition rates of handwritten words are from 42.5% [192] to 92.6% [48]. For writer-dependent cursive word recognition, the average recognition rates of seven writers are about 86% for learning files and 76% for testing files [228]. It is certain that the recognition rate can be further increased by using a larger training set.

Unlike other approaches using HMMs, MRF models are two-dimensional and they do not need to order 2-D features into 1-D space that inevitably introduces additional distortions. Comparing to the conventional relaxation labeling technique [104], MRF models do not eliminate any possible word candidates during the relaxation labeling process.

In the past decades, some impressive results in this field have been reported. Unfortunately, it is difficult to compare the performance directly between different methods, because the lack of a standard database and benchmarks: Different systems have been tested on different databases and under different conditions [99].

7

A Structural and Relational Approach to Handwritten Word Recognition

7.1 Introduction

In this chapter we present a technique that recognizes words based on structural and relational information.

The idea of describing a word in terms of its structure can be traced back to the early sixties. Eden and Halle characterized the cursive script by a set of four primitives (bar, hook, arch, loop) [77]. These primitives were detected within the word, and their orientation and location used as the basis for classification. Brown and Ganapathy extracted 183 word structural features in order to classify them *per se* [27]. They used a closed-loop pre-processing method in order to remove much of the variability in the input words. Paquet and Lecourtier used a two-stage process to recognize words in a small lexicon [192]. In the first stage, they detected significant global features, such as dominant ascending and descending strokes, and word length, in the words. Based on these features, a set of possible words was coarsely selected. The next level used precise structural information to extract finer details to differentiate between similar words. Simon also extracted the dominant features of a word by finding the difference between the word image and the axis of the word (the shortest path along the length of the word while remaining within the word itself) [228]. Using features extracted from the difference between ascending and descending strokes, the system selected possible words, which were then further examined to see if other features matched.

Essentially, a character or word can be considered as a structure that consists of parts or segments at different orientations, lengths, and positions. To analyze its structure we need to extract such oriented parts from a gray-scale image. As we discussed in the previous chapters, Gabor filters offer a solution to this. Since the Gabor filter is orientation selective, we can use a bank of n Gabor filters oriented at different angles to extract the oriented parts from words. From the oriented parts, we obtain measurements to form feature vectors which are passed to a word recognition unit for classification. In this way we can develop a recognition system that uses gray-scale images directly without any binarization, skeletonization, or size normalization operations.

7.2 Gabor Parameter Estimation

In Chapter 2 we introduced some methods for estimating Gabor parameters. Since we are to extract features directly from the gray-scale image, in this section we present several new methods for determining the parameters. Conventionally, the Gabor parameters are determined on a trial-and-error basis, in particular, the selection of the σ's. Some authors (e.g., [175, 259]) have set $\theta = c/f$, where c is a constant, usually 1. For instance, in processing texture images, many researchers use a bank containing a number of Gabor filters at various frequencies and angles. They select the upper and lower limits, and spacing of these ranges based on visual impression of the texture types in the images, or simply use a large bank of filters to cover the range of possibilities. Bovik *et al* chose the filter center frequencies depending on the type of textures encountered: for strongly oriented textures they used the most significant frequency peak along the dominant orientation, the lower fundamental frequency for periodic textures, and the two largest frequency maxima for non-oriented textures [23]. Du Buf and Heitkamper suggested that the fundamental frequency and the main orientations can be determined by analyzing the global power spectrum [76]. Casasent *et al* selected the σ parameters by assigning them to the size of a bounding rectangle [44]. In general, manual selection of the parameters is a very tedious process as pointed out by Clark *et al.* In order to make a system fully automatic it is necessary that the selection of the parameters be data driven [57].

Due to large variations present in word size and thickness as shown in Figure 7.1, assigning constant values to the parameters is unsuitable to handwritten word recognition. The correct selection of the parameters is essential to extracting a consistent and meaningful set of parts from a range of word images, which vary significantly. The effects of incorrect parameter selection are shown in Figure 7.2 for the word image shown in Figure 7.2(a). Figures 7.1(c) and (d) show the effect of incorrect parameter selection, where the parameters used are the same as those used for the word in Figure 7.2(a). Figures 7.1(e) and (f) show the results with the parameters correctly selected. The values of the parameters f, σ_x, and σ_y are determined by the properties of the word image - the size of the characters and the width of the writing.

In this section, we describe how the parameters $\Delta\phi$, θ, f, σ_x, and σ_y of the Gabor filter are estimated. To determine relationships between the measurable properties and the Gabor filter parameters, we derive the relationships between the parameters and the length and width of line segments.

Estimation of $\Delta\phi$

Since the the number of Gabor angles (ϕ), or their spacing $\Delta\phi$ determines the number of Gabor filters for extracting the oriented parts and does not

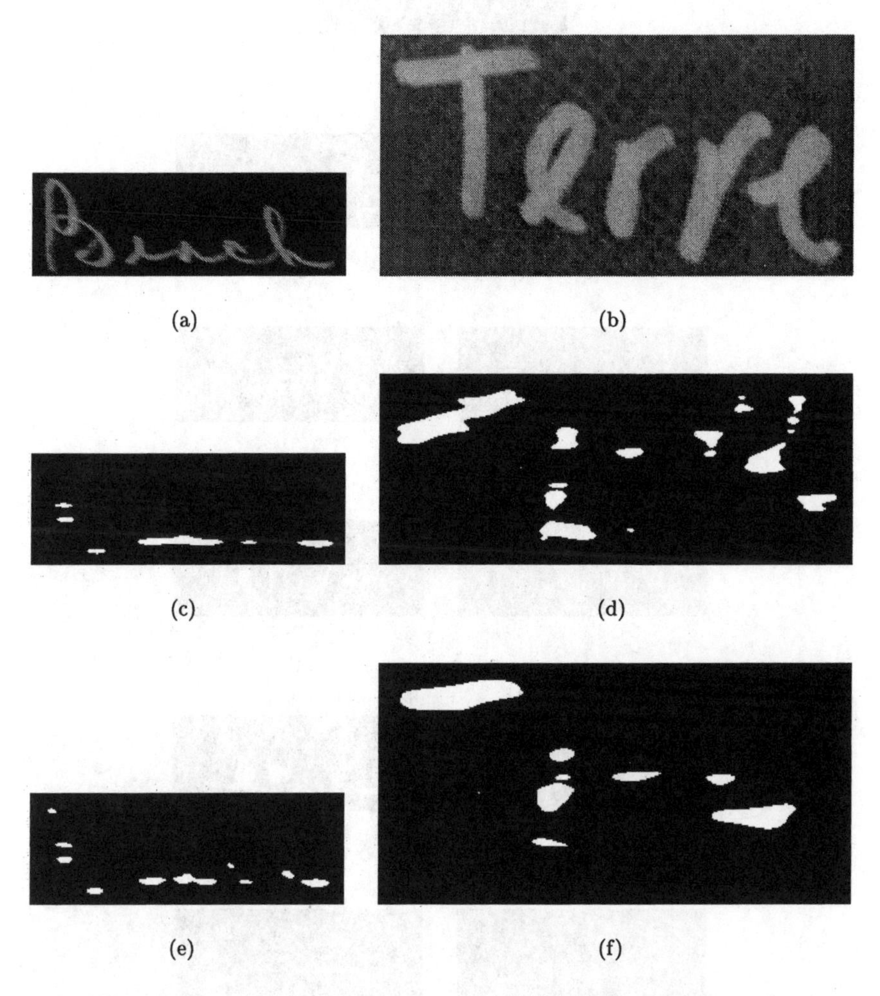

FIGURE 7.1. Effects of maintaining the same Gabor filter parameters, for parts oriented at 0°, for two different words. (a) image of word *Beach*, (b) image of word *Terre*, (c,d) extracted parts with incorrect parameter selection ($\sigma_x = 0.0445$, $\sigma_y = 0.0159$, $f = 50$, and $\theta = \phi + 90°$), (e,f) extracted parts with correct parameter selection ($\sigma_x = 0.0251$, $\sigma_y = 0.0159$, $f = 51.2$, and $\theta = \phi + 90°$), ($\sigma_x = 0.1004$, $\sigma_y = 0.0050$, $f = 16.0$, and $\theta = \phi + 90°$).

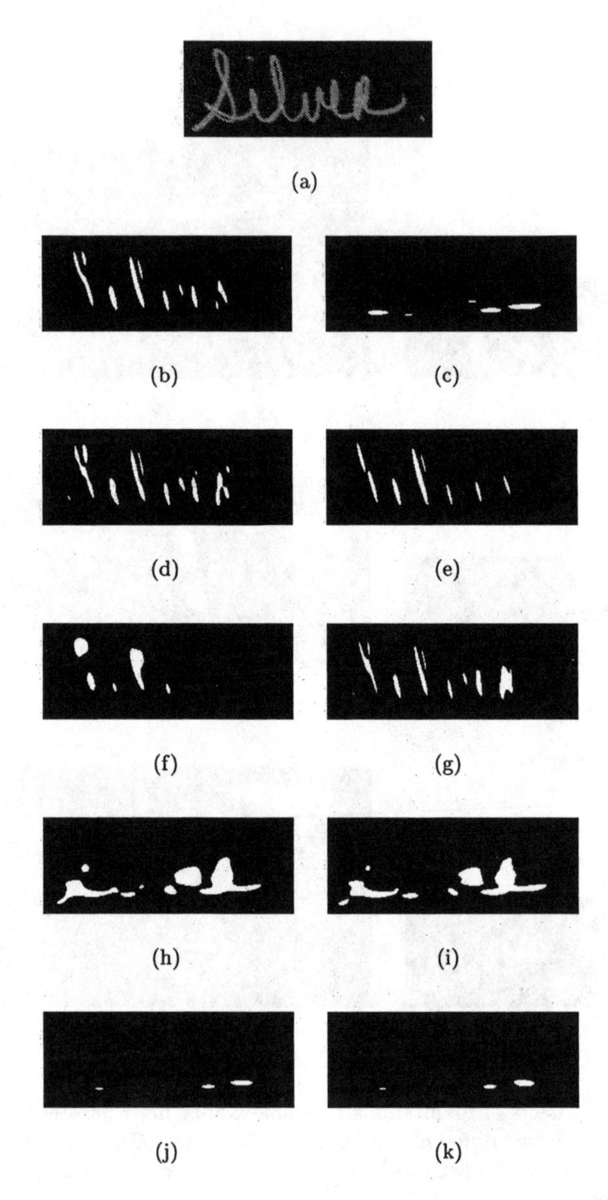

FIGURE 7.2. Effect of the incorrect parameter selection. (a) Gray scale reference word, (b) correct parameter selection $f = 50$, $\sigma_x = .05$, $\sigma_y = 0.015$, $\phi = 105°$. (c) correct parameter selection $\phi = 0°$, (d-g) variation of σ_x and σ_y with $f = 50$ and $\phi = 105°$, (i-k) variation of f with $\sigma_x = .05$, $\sigma_y = 0.015$. (d) $\sigma_x = 0.025, \sigma_y = 0.015$, (e) $\sigma_x = 0.1, \sigma_y = 0.015$, (f) $\sigma_x = 0.05, \sigma_y = 0.03$, (g) $\sigma_x = 0.05, \sigma_y = 0.0075$, (h) $f = 10$, (i) $f = 30$, (j) $f = 70$, (k) $f = 90$.

contribute to the optimization of the shape of the Gabor filter, we can determine $\Delta\phi$ by examining the reconstruction of the extracted parts. For the word image shown in Figure 7.2(a), Figure 7.3 shows the reconstruction of the parts for $\Delta\phi = 22.5°$, $\Delta\phi = 15°$, $\Delta\phi = 11.25°$, $\Delta\phi = 5.625°$. The reconstructed word was created by ORing together the extracted parts from the n views, which is a typical example with other word images producing similar results. As shown in Figure 7.3, the reconstruction for $\Delta\phi = 22.5°$ shows that insufficient parts of the word are extracted (parts of the l and S are missing). The reconstruction for $\Delta\phi = 15°$, $\Delta\phi = 11.25°$, and $\Delta\phi = 5.625°$ are virtually identical with the important parts of the words being extracted. In general, for handwritten English words, $\Delta\phi = 15°$ (12 angle bands in the range [0°, 180°]) is sufficient.

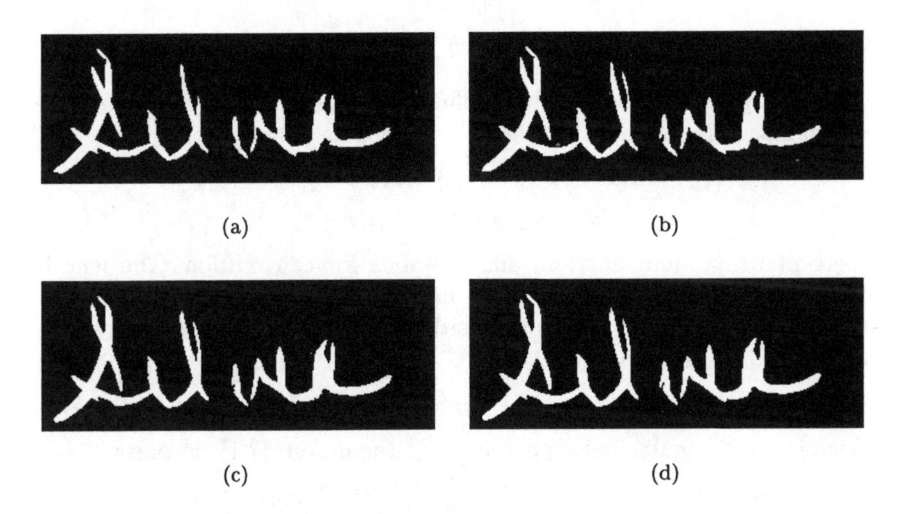

FIGURE 7.3. Reconstruction of the parts of the word in Figure 7.11(b) for a change in the Gabor angle of (a) $\Delta\phi = 22.5°$ (8 angles), (b) $\Delta\phi = 15°$ (12 angles), (c) $\Delta\phi = 11.25°$ (16 angles), (d) $\Delta\phi = 5.625°$ (32 angles).

Estimation of θ

We set the value of θ at $\phi + 90°$, which maximizes the effect of the Gabor filter in the direction of the filter (ϕ). The importance of this will become clear in the following sections when estimating the parameters f and σ_y. Figure 7.4 gives an example of a Gabor filter with $\phi = 0°$ and $\theta = 90°$.

Estimation of f

The response of applying a Gabor filter to a line is given by

$$r(x,y) = g(x,y) * p(x,y) \tag{7.1}$$

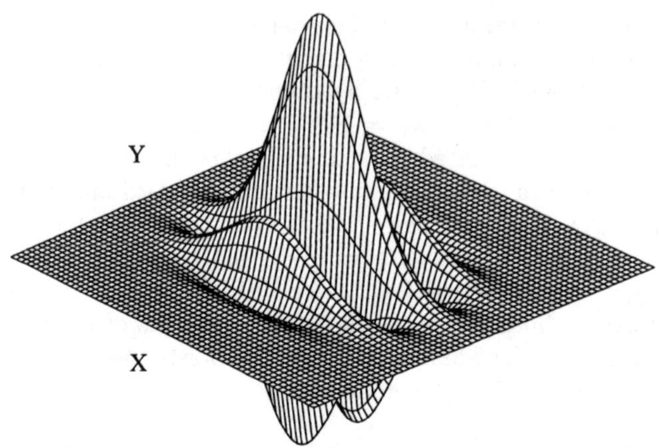

FIGURE 7.4. Spatial representation of a Gabor filter with $\theta = 90°$ and $\phi = 0°$.

where $g(x, y)$ is the Gabor filter defined by (3.16), and $p(x, y)$ is a 2-D rectangular pulse oriented at 0° can expressed as follows:

$$p(x, y) = \mu(x - \tau_{x1}, y - \tau_{y1}) - \mu(x - \tau_{x1}, y - \tau_{y2}) \\ -\mu(x - \tau_{x2}, y - \tau_{y1}) + \mu(x - \tau_{x2}, y - \tau_{y2}), \quad (7.2)$$

where $\mu(\cdot)$ is a step function, and $*$ stands for convolution. The length and width of the pulse are $\tau_{x2} - \tau_{x1}$ and $\tau_{y2} - \tau_{y1}$, respectively. (7.2) is a rectangular pulse at an arbitrary position. As the Gabor filter is invariant to position, we can write (7.2) as

$$p(x, y) = \mu(x, y) - \mu(x, y - \tau_y) - \mu(x - \tau_x, y) + \mu(x - \tau_x, y - \tau_y) \quad (7.3)$$

which is a rectangular pulse positioned at the origin. (7.1) becomes

$$r(x, y) = \int_{-\infty}^{y} \int_{-\infty}^{x} g(\alpha, \beta)\, \mathrm{d}\alpha\, \mathrm{d}\beta - \int_{-\infty}^{y-\tau_y} \int_{-\infty}^{x} g(\alpha, \beta)\, \mathrm{d}\alpha\, \mathrm{d}\beta - \\ \int_{-\infty}^{y} \int_{-\infty}^{x-\tau_x} g(\alpha, \beta)\, \mathrm{d}\alpha\, \mathrm{d}\beta + \int_{-\infty}^{y-\tau_y} \int_{-\infty}^{x-\tau_x} g(\alpha, \beta)\, \mathrm{d}\alpha\, \mathrm{d}\beta \quad (7.4)$$

where α and β are dummy variables.

Unfortunately, there is no closed form solution to (7.4), as the solution of a Gabor step response does not exist [24]. The response of the Gabor filter to the pulse (7.2) will be of the form shown in Figure 7.5 which shows the response along a column through the middle of the rectangular pulse. The dotted line A-B represents the width of the pulse and its position.

However, the frequency of the response will remain essentially the same as the frequency of the Gabor filter. As the frequency decreases, as shown in Figure 7.5, the distance between the two response peaks decreases until a frequency, at which they overlap, has been reached. This is the frequency that produces the maximum response with a single peak, and is given by

$$f = \frac{N}{2W} \quad (7.5)$$

where N is the size of the image and W is the width of the line.

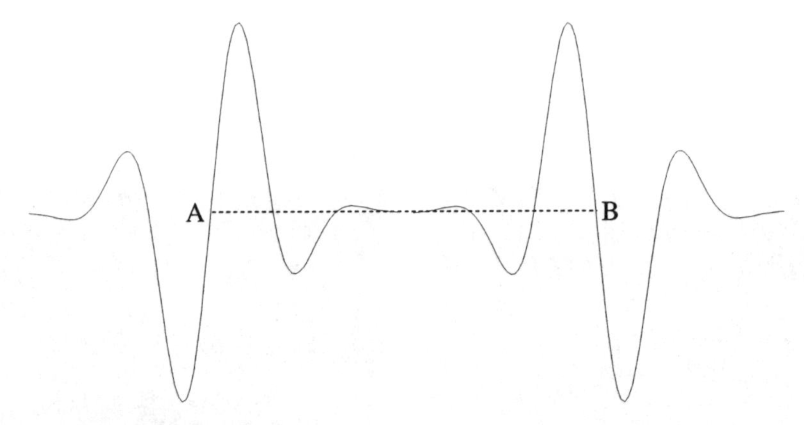

FIGURE 7.5. Response of a Gabor filter to a line AB. Note that the response occurs at the points of change.

Estimation of σ_x

The parameter σ_x determines the spread of the Gabor filter in the ϕ direction, or along the x axis when $\phi = 0°$, and thus determines the minimum length of the part which can be extracted. We can observe this in Figure 7.6 which shows the result of a Gabor filter with a set of σ_x values applied to an image containing five horizontal lines of different lengths. The direction of the Gabor filter is $\phi = 90°$ and the value of σ_y is selected to match the width of the line (10 pixels). As the value of σ_x increases, the responses to the shorter lines decrease until they disappear. This is due to the fact that the spectrum of the shorter lines is larger (in the u direction) than the spectrum of the Gabor filter in the same direction. When the spectrum of the Gabor filter matches, or is greater than the spectrum of the line, the Gabor filter will produce the maximum response.

For the words shown in Figures 7.11 and 7.12, the minimum length of the parts to be extracted should be determined by the height of the central part of the smaller characters in the word (e.g., characters *i, v, e, r* in Figure 7.11). The heights of the smaller characters in the word are measured, and then their average height H' is calculated. We measure the heights of the smaller characters by segmenting the word into vertical parts (as performed in Section 7.3.1, Figure 7.8(d)), measuring the heights of these segmented areas, rejecting the taller segments, then averaging them. However, we have to stress that this process is not character segmentation as there is no attempt to separate the characters, only to extract the vertical parts of the characters. As these smaller characters consist of curved or smaller parts, we set $H = 0.75H'$, the height value used in determining σ_x.

From Figure 7.6, it is clear that as the line becomes shorter, a smaller value of σ_x is needed to detect it. As can be seen from Figure 2.13(d), which shows the magnitude of the frequency domain, when the value of σ_x increases, the spread in the frequency domain decreases, and similarly

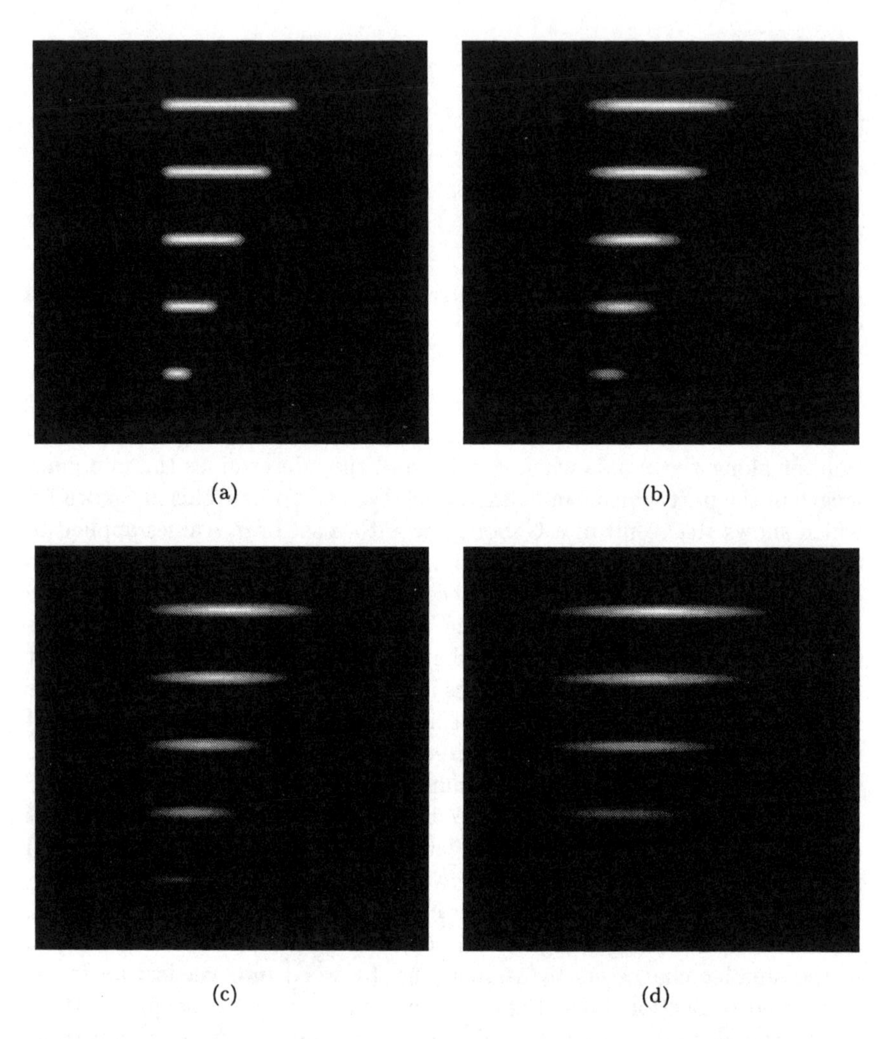

FIGURE 7.6. Effect of a change in the parameters σ_x of the Gabor filter with $\phi = 0°$, $f = 50$, $\sigma_y = 0.015$. The images show the power response for five horizontal lines with lengths, from top to bottom, 100, 80, 60, 40, 20 pixels. (a) $\sigma_x = 0.02$, (b) $\sigma_x = 0.05$, (c) $\sigma_x = 0.11$, (d) $\sigma_x = 0.2$.

when σ_x decreases, the spread in the frequency domain increases. This is an example of the *Similarity theorem of Fourier transforms* defined as follows [216]: if $\mathcal{F}\{g(x,y)\} = G(f_x, f_y)$, then $\mathcal{F}\{g(ax,by)\} = \frac{1}{|ab|}G\left(\frac{f_x}{a}, \frac{f_y}{b}\right)$. This is consistent with the spread in the frequency domain of a rectangular box. The value of σ_x can be approximated by equating the spread of the Gabor spectrum to the spread of a rectangular box's spectrum at their half height points. The spread of a rectangular box is a 2D sinc function [98]:

$$|F(u,v)| = AXY\left|\frac{\sin(\pi uX)}{\pi uX}\right|\left|\frac{\sin(\pi vY)}{\pi vY}\right|,$$

where A is the amplitude, X is the length of the rectangle in the x direction, and Y is the length of the rectangle in the y direction. When we consider the variation along the u axis only ($v = 0$) and normalize AXY to 1, from the above we have

$$|F(u,0)| = \left|\frac{\sin(\pi uX)}{\pi uX}\right|.$$

The frequency at the half height point, normalized to the image size is therefore

$$u = \frac{0.603355N}{X}. \tag{7.6}$$

Similarly for the Gabor frequency response (3.16), considering the variation along the u axis, ($v = v_0, \phi = 0°$) becomes

$$G(u, v_0) = \exp\left(-\pi u^2 \sigma_x^2\right)$$

with $u_0 = 0$ as $\theta = 90°$. The frequency at the half-height point is

$$u = \frac{1}{\sigma_x}\sqrt{\frac{ln2}{\pi}}. \tag{7.7}$$

Equating (7.6) and (7.7) gives

$$\sigma_x = \frac{H}{aN}\sqrt{\frac{ln2}{\pi}} \tag{7.8}$$

where $H = X$ and $a = 0.603355$ which is obtained from the half height of the sinc function.

Estimation of σ_y

The parameter σ_y determines the spread of the Gabor filter in the $\phi + 90°$ direction, or along the y axis when $\phi = 0°$. As the width of the line increases, the value of the required σ_y decreases. In the frequency domain, a decreasing σ_y results in a larger spectral spread. This is opposite to the spectral spread of a line (as the line width increases, its spectral spread decreases). This is the consequence of the interactions between the responses at the edges of the line (Figure 7.5) similar to that used for determining the frequency.

The value of the parameter σ_y is determined by the width of the strokes within the word. As Figure 7.5 shows, the responses occur at the top and bottom edges of the rectangular pulse. This was the main motivation behind the selection of θ as $\phi + 90°$. When the frequency is correctly selected, the resulting response will be a single peak along the center of the pulse. As there is no closed form solution to (7.1), we have to find a relationship between σ_y and the width of the pulse W with numerical means. By examining the results of the parts extracted over an extensive range of combinations of σ_y's and f's, we found the relationship to be

$$\sigma_y \approx \frac{1}{4\pi W} \tag{7.9}$$

where W is the width of the rectangular pulse. The relationship is determined *intuitively* from (7.5), as the same mechanism of the interactions between the two edge responses, which determines the relationship for f and σ_y. We can also write (7.9) as $\sigma_y = f/2\pi N$ (by substituting (7.5) into (7.9)), which shows a linear relationship between σ_y and f.

7.3 Feature Extraction

As alluded to in Section 7.1, we can consider a character, and thus a word as consisting of a set of oriented parts. The set of n Gabor filters are used to extract these oriented parts from the gray-scale word image. The filtered outputs are then processed to obtain binary images of the extracted parts that are used for calculating unary and binary features. Using these features, the recognition system can then perform classification. Figure 7.7 shows a block diagram of the system. Ho *et al* used a similar concept when they extracted the parts of a binarized word, although only in four directions, by detecting continuous runs of pixels along those directions [109]. These are used as the features in the image.

As discussed previously, in handwriting recognition, we need to reduce variations in data as much as possible by performing pre-processing, e.g., slop and slant correction, and normalization. In the following we present some additional pre-processing techniques based on structural and relational information, which we will use in this chapter.

7.3.1 Slant Correction

Slant correction is first performed on word images to remove variations in the slope of writing. The corrected words will produce a consistent set

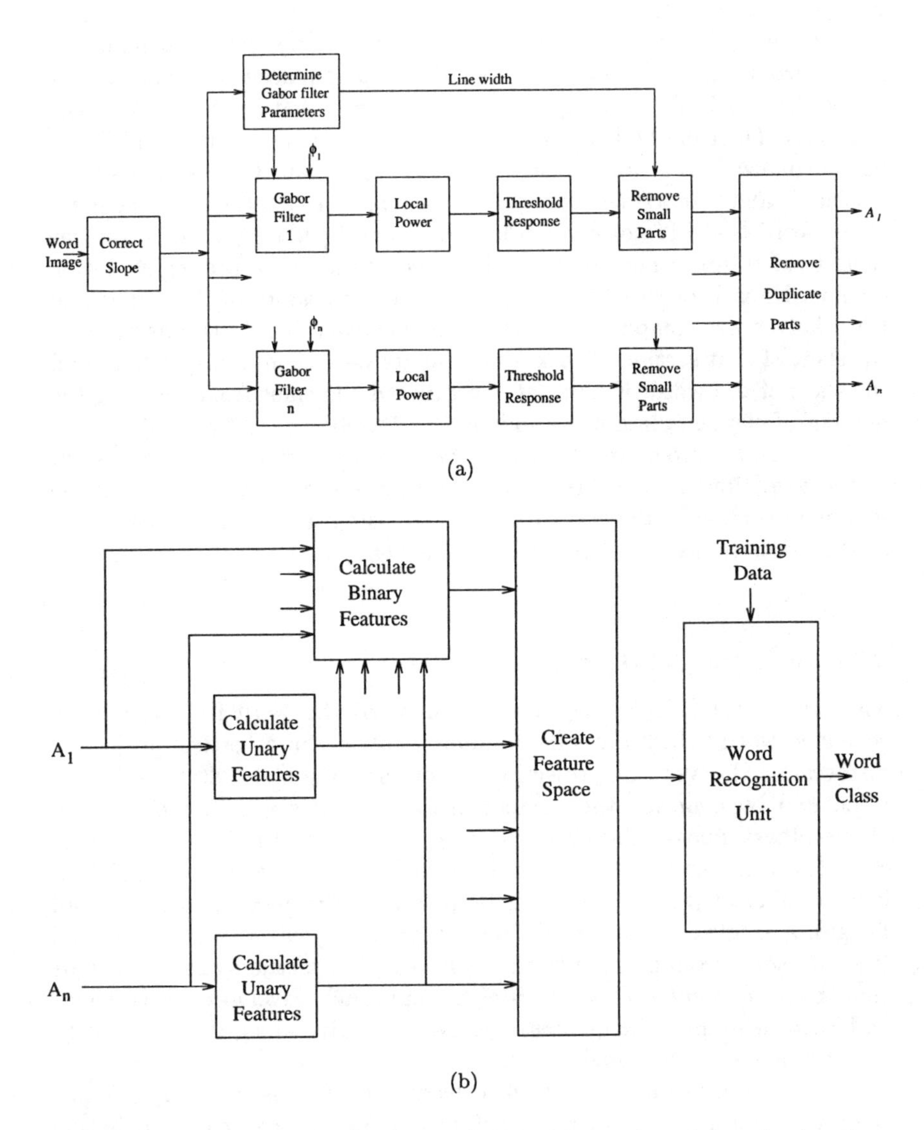

FIGURE 7.7. Block diagram for the word recognition system, (a) part extraction, (b) feature extraction and word recognition.

of features over the range of words used by the system. The underlying assumption in virtually all methods used to calculate the slant is that words are normally written with a *consistent* slant, thus the global slant of the word can be calculated.

Brown and Ganapathy used the angle between the crossover points of two horizontal lines displaced by a small distance placed through the center of the word. The average of the angles was used as the slant of the word [27]. Their method requires thinned images that have been first baseline normalized, and words that must be greater than two characters long. Bozinovic and Srihari used binarized word images to analyze only the parts of the word that did not contain long horizontal parts [26]. The word was then divided into a number of sections determined by a pre-segmentation scheme. In each of these sections, the center of gravity of the upper and lower halves was computed, and the angle between them was considered as the slant of that section. The sectional slants were then averaged to obtain the slant of the word. Kimura *et al* estimated the slant from counting the number of elements in a chain code at 0° 45°, 90°, and 135° [138].

We introduce two more effective methods for determining the global slant of the word image: The first method estimates the slant by finding the minimum entropy in the vertical projection histograms, whereas the second method determines the slant from the frequency domain.

Minimum Entropy of Distribution

The first method finds the dominant slant of the word from the slant-corrected word, which gives the minimum entropy of a vertical projection histogram. The vertical projection histogram (Figures 7.8(c) and(d)) is calculated by counting the number of foreground pixels in each column of the binary image. The distribution is then normalized to a total area = 1. The gray-scale image is first binarized using the method in [187], as it is suitable for the type of distributions found in most gray-scale word images: A single dominant peak followed by a long tail. Figures 7.8(a) and (b) show some examples of binarized images. This method was found to be similar to the *slanted vertical histogram* approach developed by Guillevic and Suen, who used the greatest positive derivative along the distribution to determine the slant angle [102].

We can describe our basic idea by considering a line as an example. When the line is slanted at an angle, it will have a low wide spread distribution whose width is $l \cos \beta$, where l is the length of the line and β is the is angle of the line to the horizontal axis. When the line is upright, the distribution will be tall and narrow, which will result in a lower entropy measure than that of the low wide spread distribution of the slanted line. Thus the entropy can be used as a measure of the uprightness of the word.

We calculate the vertical projection histogram by first correcting the binary image by an arbitrary angle α_i, using

$$x' = x - y \tan \alpha_i,$$
$$y' = y, \tag{7.10}$$

where α_i is measured relative to the norm, and (x, y) are the coordinates of a pixel in the binary image. A new vertical projection histogram is then formed for each new skewed image. The entropy of the vertical projection histogram is calculated by

$$H_\alpha = -\sum_{j=1}^{N} p_j \log p_j,$$

where N is the number of columns in the histogram and p_j is the probability of a foreground pixel appearing in column j.

The range R of values used for α_i is $[-60°, 60°]$. This range adequately covers the writing styles encountered. The correction angle α_m is found from the minimum entropy, i.e.,

$$\alpha_m = \min_{\alpha \in R} H_\alpha.$$

Using (7.10) and α_m we can correct the slant of the gray-scale word image by relocating a pixel at (x, y) to a new location (x', y') resulting in a slant-corrected image (e.g., Figure 7.8(b)). The vertical projection histograms can be efficiently generated by maintaining a table of the locations of the foreground binary pixels (e.g., from Figure 7.8(a)), then manipulating only these points in generating the vertical projection histograms.

Figure 7.8(c) shows the vertical projection histogram for the slanted word in Figure 7.8(a), and Figure 7.8(d) shows the vertical projection histogram for the same word corrected by α_m (Figure 7.8(b)). A comparison of this histogram to the one in Figure 7.8(c) indicates that it is far more compact and upright, which is consistent with the analogy of the vertical line. A disadvantage of this method is that the slant of a large letter may tend to dominate the calculation as it contains a greater area of foreground pixels, but assuming the slant is *consistent*, its effect is insignificant.

Frequency Domain

We present a frequency-domain approach to calculating the slant of the word image, which is based on the method proposed in [7]. In this approach, the word image is first transformed into the frequency domain, then the magnitude image $M(u, v)$ is calculated. Figure 7.9 shows the spectral magnitude for the word image shown in Figure 7.11(a). We use a well-known property of the Fourier transform that the directionality of the image is

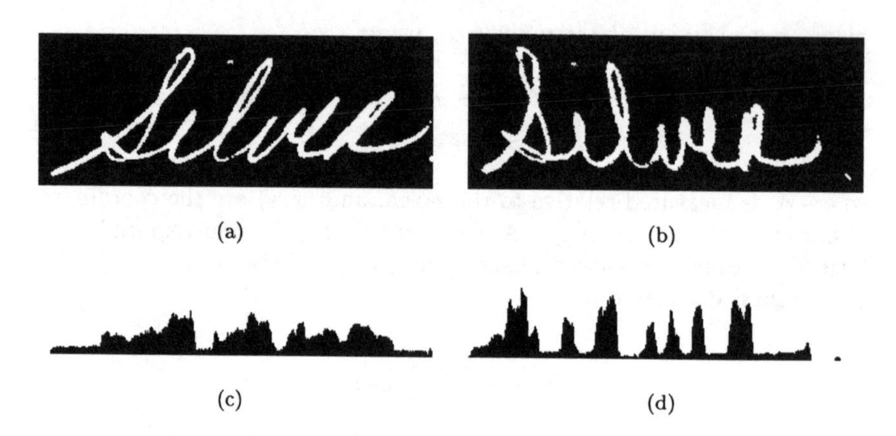

FIGURE 7.8. Slant correction using vertical projection histograms. (a) binary image of uncorrected word, (b) binary image of slant-corrected word, (c,d) vertical projection histograms.

preserved in the frequency domain. Rewrite (3.16) and (3.16) as follows:

$$\begin{aligned} g(x,y) = & \exp\{(a\cos\phi + b\sin\phi)^2 + (c\cos\phi + d\sin\phi)^2\} \times \\ & \exp\{\jmath 2\pi(u_0 x + v_0 y)\}, \\ G(u,v) = & \exp\{(e\cos\phi + f\sin\phi)^2 + (g\cos\phi + h\sin\phi)^2\}, \end{aligned}$$

where a, b, c, d, e, f, g, h are substitutes for the remainder of the equations. The direction ϕ has been maintained in both domains. Figures 2.12(c) and 2.13(c) illustrate this property: The directions of the Gaussian envelopes have been both rotated by 45°. Fourier transform extracts *global* properties from the image. It is the global property that enables us to determine the global slant of the word. Figure 7.9 shows the global orientation, noting that the direction in the frequency domain is rotated by 90° from that in the spatial domain.

To extract the directional information from the magnitude image $M(u,v)$, we express the image in the polar coordinate system (ρ,θ). We use bilinear interpolation to construct the polar image $M(\rho,\theta)$ from that in the Cartesian coordinate system [216]. For each of the ρ columns, we find the maximum value θ and construct a histogram H_θ of θ_{max} as shown in Figure 7.10 for the spectral magnitude image shown in Figure 7.9. The slant of the word is then calculated from the weighted average

$$\alpha = \frac{\sum\limits_{\beta\in S}\beta w_\beta}{\sum\limits_{\beta\in S} w_\beta} \tag{7.11}$$

where S is the set of angle values γ in H_θ such that $H_\theta(\gamma) \geq 0.5 \max H_\theta$, $\gamma \neq 90°$, and the support of S is contained in the range $[30°, 150°]$. The range $[30°, 150°]$ is the same as that for the other methods in order to cover all writing styles and to exclude horizontal strokes, the condition, $\gamma \neq 90°$, removes the artifact (along the u axis) due to the vertical components of the word.

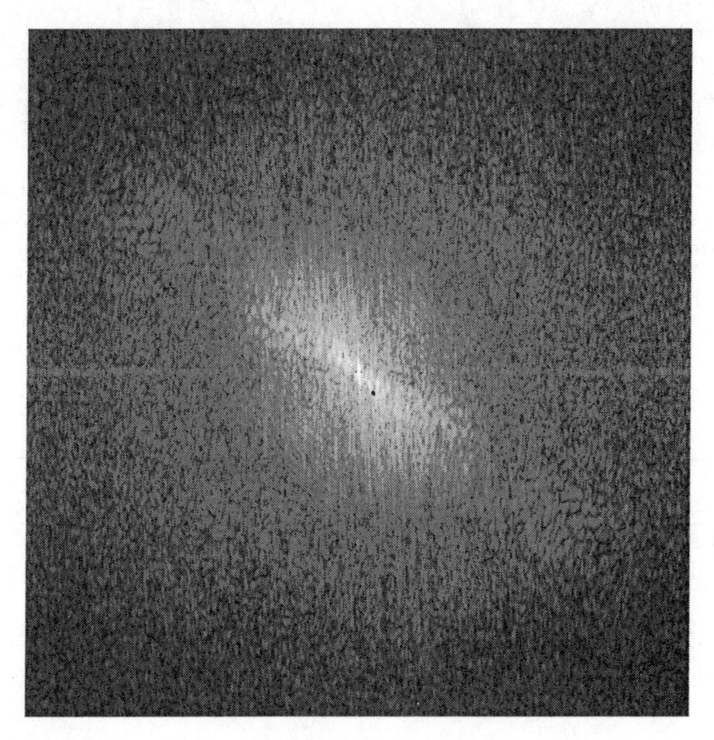

FIGURE 7.9. Spectral magnitude (displayed as $\log(p+1)$) for the word image shown in Figure 7.11(a).

Slant Correction Results

Table 7.1 gives the results for the slant estimation methods. From these results, we select the Fourier domain method for estimating the slant of words as it gives the best, visually consistent results. The Minimum Entropy method was the quickest of the two. Figure 7.11(b) shows the slant-corrected word in Figure 7.11(a). Figure 7.12 shows some more examples of the slant correction for words with positive, negative, and zero slants. The Fourier domain method uses the gray-scale image directly as opposed to the Minimum Entropy method which first binarizes the image. The use of the gray-scale images is advantageous as it removes one level of processing and reduces information loss and distortion.

FIGURE 7.10. Histogram of θ_{max} for the spectral magnitude image in Figure 7.9. The horizontal axis is from 0° to 179°. Note that the histogram angles have been rotated by 90°.

Method	Word				
	Silver	Beach	Terre	Ronks	Baton
Minimum Entropy	51°	34°	63°	121°	93°
Frequency domain	52°	34°	71°	123°	92°

TABLE 7.1. Calculated slant angles, α_m, for the words shown in Figures 7.11(a), 7.12(a), 7.12(c), 7.12(e), 7.12(g). The tilt angles are specified relative to the x-axis.

FIGURE 7.11. Gray-scale image of word *Silver*: (a) the slant angle, (b) the original image, (c) the slant-corrected image.

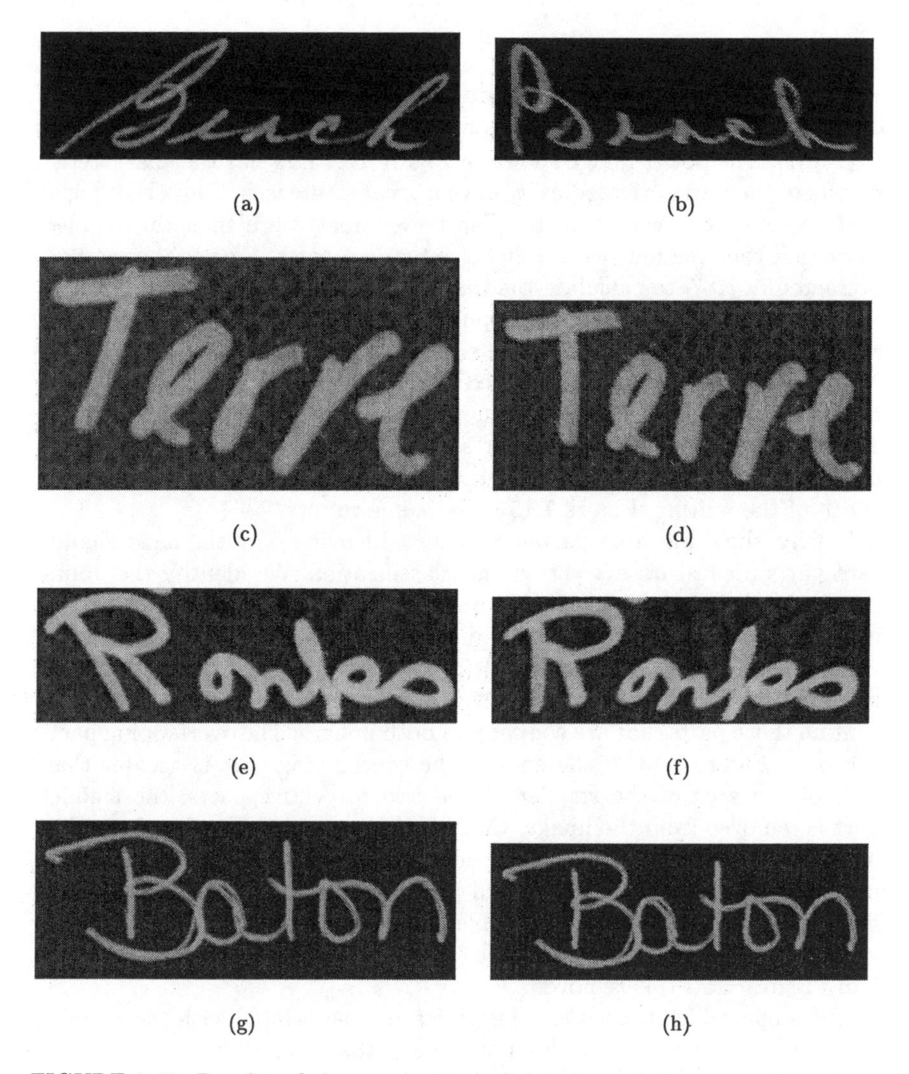

FIGURE 7.12. Results of slant correction. Original word images on left, slant corrected word images on right.

7.3.2 Part Extraction

The word image is first applied to the bank of n Gabor filters, written as

$$r_i(x,y) = w(x,y) * g_i(x,y) \qquad i = 1,2,\ldots,n \tag{7.12}$$

$w(x,y)$ is the word image, $g_i(x,y)$ is the Gabor filter oriented at ϕ_i, and $*$ represents the convolution operation. The output of Eq. (7.12) produces a complex response. From $r_i(x,y)$, we calculate its power $P_i(x,y) = |r_i(x,y)|^2$. The power image shown in Figure 7.14 now has the characteristic where the parts oriented at ϕ_i have a greater intensity than other parts that are oriented away from ϕ_i. The power images are then thresholded to extract the oriented (with a higher intensity) parts. We use the method presented in [187] to calculate the thresholding point of the power image.

The next stage is to filter out spurious parts in the binary image thus retaining only those parts that are significant to the reconstruction of the word. The size of the parts removed must depend on the stroke width of the writing, because words written with thick pens (e.g., Figure 7.1(b)) will have much larger spurious parts than that with thin pens (e.g., Figure 7.1(a)). The minimum area of the parts was chosen to be the squared width of the writing. Figure 7.15 shows some results.

Finally, the duplicated parts are removed leaving only the most significant parts for feature extraction and classification. We identify the duplicated parts by examining the parts present in neighboring images which are images generated by *neighboring* Gabor filters that are oriented at adjacent angles, e.g., Figures 7.15(a)-(b), (b)-(c), ..., and (l)-(a), are neighboring images. When the two images are ANDed together the resulting image will contain those parts that are common to both images: The overlapping parts shown in Figure 7.13. If the area of the overlapping part is greater than 75% of the area of the smaller of the two respective parts, the smaller part is removed from the image. Consider an example, the parts from Figure 7.13(a) (the overlapping areas between Figures 7.15(c) and (d)): The areas of these parts and the ANDed parts are shown in Table 7.2. For the first part on the left (part 1 in Table 7.2), the area of the ANDed part is greater than 75% of the area of the smaller part (241), so that part (part 1 of Figure 7.15(c)) is removed.

The angle, which is used as a unary feature, associated with the remaining part is assigned the weighted average of the two parts:

$$\phi_a = \frac{w_1\phi_1 + w_2\phi_2}{w_1 + w_2}$$

where w_1 and w_2 are the areas of each of the respective parts. In this way we can remove smaller, less significant parts. This will aid the classification, as only one part of each portion of a character is used in classification.

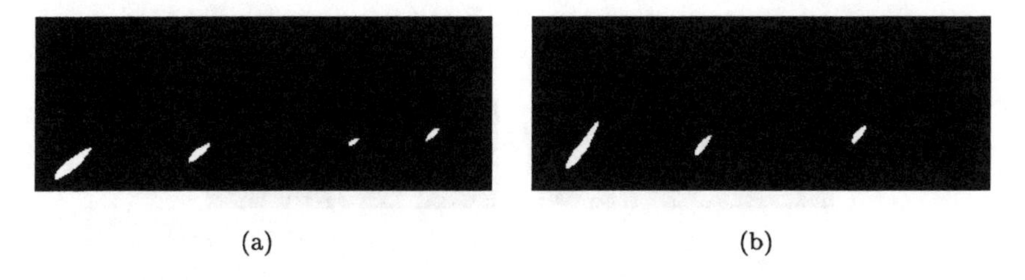

(a) (b)

FIGURE 7.13. Extraction of overlapping parts. (a) ANDed images of Figure 7.15(c) and (d), (b) ANDed images of Figure 7.15(d) and (e).

	Area of Part			Area of Part		
Part	Image 1	Image 2	ANDed	Image 1	Image 2	ANDed
1	241	322	195	322	405	209
2	69			128	105	71
3	109	128	87	75	118	59
4	99	75	23	48	46	
5	156	48	43			

TABLE 7.2. Areas of the extracted parts. Left hand pair of images obtained from 7.15(c) and (d), right hand pair of images obtained from 7.15(d) and (e). The parts are numbered from the left.

7.3.3 Feature Extraction

Once the oriented parts have been extracted from the word, e.g., Figure 7.16, they are used to generate the unary and binary features. Unary features are features that are confined to a single part, e.g., the area of the part. Binary features, on the other hand, are features that describe the relationship between two parts, e.g., the distance between the centroids of the two parts. Thus each part possesses a set of unary features and each pair of parts possesses a set of binary features. The selected unary features characterize each of the parts in size and position. The n processed outputs in conjunction with the unary features are then used to calculate the binary features which indicate how the individual parts are interrelated with each other.

Unary features

For the extracted parts as that shown in Figure 7.16, we calculate the following unary features for each part in each of the n images:

1. Direction of the Gabor filter ϕ or ϕ_a in the case where duplicate parts have been removed. This will represent closely the angle of each of the extracted parts.

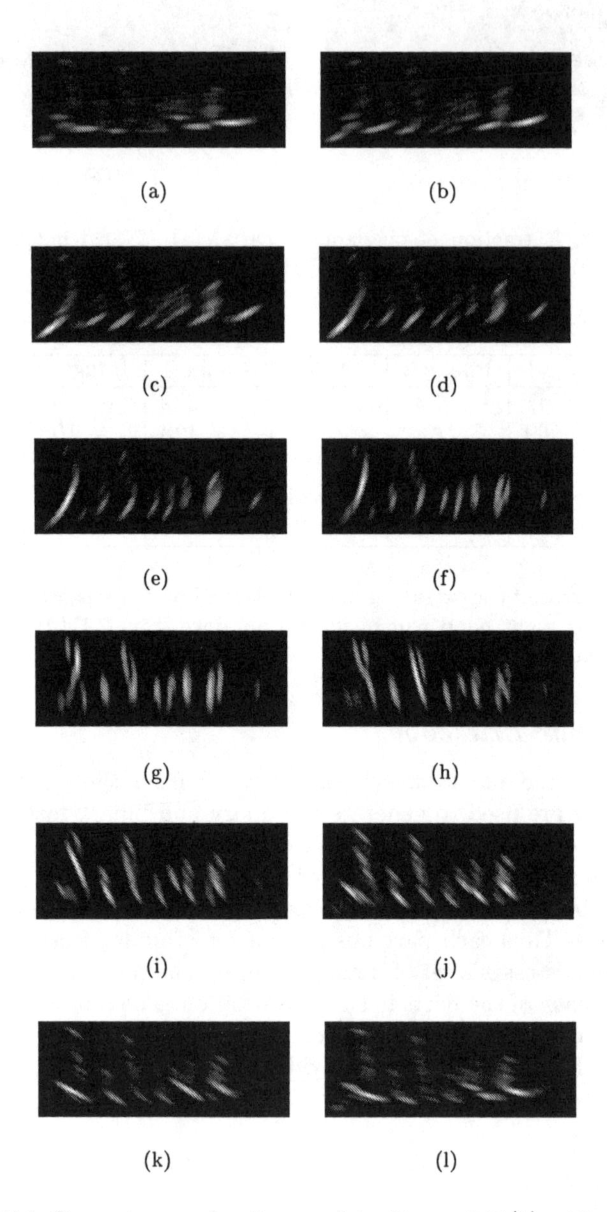

FIGURE 7.14. Power images for the word in Figure 7.11(b) with $\sigma_x = 0.05$, $\sigma_y = 0.015$, $f = 50$, $\theta = \phi + 90°$, and $\Delta\phi = 15°$. Angle of Gabor filter, ϕ: (a) 0°, (b) 15°, (c) 30°, (d) 45°, (e) 60°, (f) 75°, (g) 90°, (h) 105°, (i) 120°, (j) 135°, (k) 150°, (l) 165°.

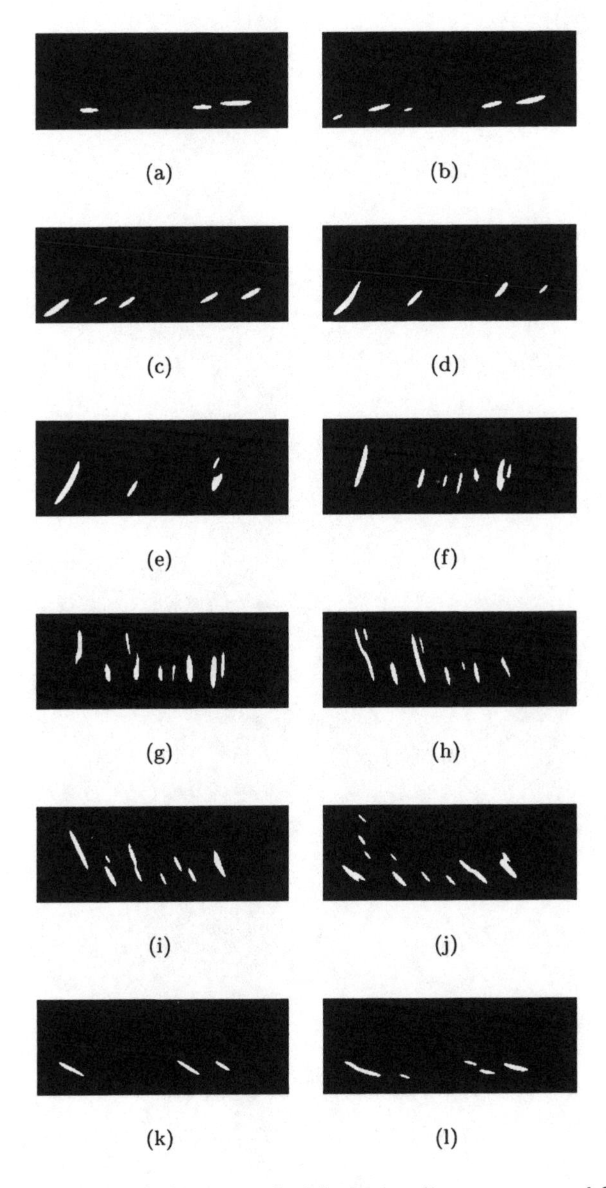

FIGURE 7.15. Thresholded images of with the small parts removed for the word in Figure 7.11(b) with $\sigma_x = 0.05$, $\sigma_y = 0.015$, $f = 50$, $\theta = \phi + 90°$, and $\Delta\phi = 15°$. Angle of Gabor filter, ϕ: (a) 0°, (b) 15°, (c) 30°, (d) 45°, (e) 60°, (f) 75°, (g) 90°, (h) 105°, (i) 120°, (j) 135°, (k) 150°, (l) 165°.

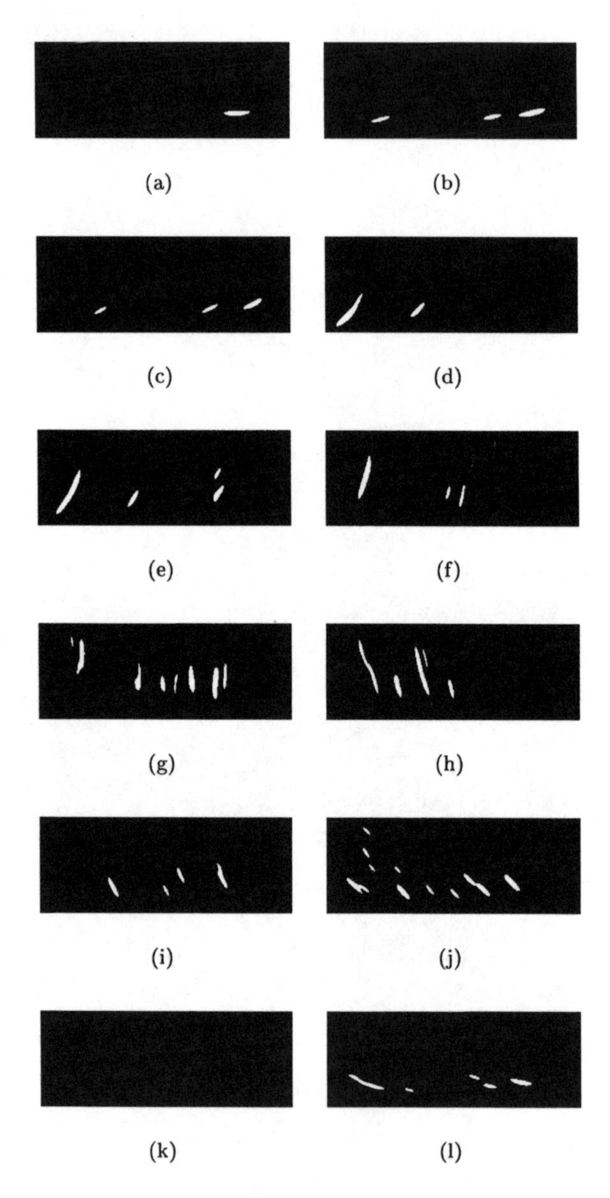

(a) (b) (c) (d) (e) (f) (g) (h) (i) (j) (k) (l)

FIGURE 7.16. Extraction of oriented parts, dominant parts from Figure 7.15 from the word in Figure 7.11(b) with $\sigma_x = 0.05$, $\sigma_y = 0.015$, $f = 50$, $\theta = \phi + 90°$, and $\Delta\phi = 15°$. Angle of Gabor filter, ϕ: (a) 0°, (b) 15°, (c) 30°, (d) 45°, (e) 60°, (f) 75°, (g) 90°, (h) 105°, (i) 120°, (j) 135°, (k) 150°, (l) 165°.

2. Normalized area of each part:

$$A_n = \frac{A_i}{\sum A_i}, \qquad i = 1, 2, \ldots, N$$

where N is the total number of parts in all of the n images. The normalized area is selected as a feature that is invariant to the thickness of the strokes in the word, assuming that the thickness is consistent over the whole word. If two identical words are written, one with a thick pen and the other with a thin pen, the value of the normalized area for each part of the two words will be the same. Also, if the two words are written in different sizes, the normalized area will be the same for the two words.

3. Normalized position of y centroids:

$$Y_n = \frac{y - y_{min}}{y_{max} - y_{min}}$$

where the values of y_{max}, y_{min} are found from the parts contained in all of the images. The position of the y centroids are normalized in the range [0,1] to the height of the word, providing height invariance. Thus, the position of the parts in the words of different sizes are scaled to the same size to aid identification.

4. Normalized position of x centroids:

$$X_n = \frac{x - x_{min}}{y_{max} - y_{min}}$$

where the values of x_{min}, y_{max}, y_{min} are found from the parts contained in all of the images. The position of the x centroids are normalized to the height of the word. This allows words of different lengths to be easily distinguished, as the height/length ratio of words should be relatively consistent for each type of words. The relative position of the centroids are referenced to the leftmost and lowest positions of the part centroids. This allows the word to be positioned anywhere within the image.

The normalization of the features results in the features being invariant to size and position of the word within the image. It is performed only on the extracted feature points, and not on the image thus avoiding degradation.

Binary features

Binary relationships between parts are determined between pairs of parts. The two parts can be from any of the n images that have a distance between their centroids of at least d_b. For each of the pairs which meets this criterion, we generate the following binary features:

1. Relative overlap:

$$RO = \frac{A_o}{A_i + A_j - A_o}$$

where A_o is the area overlapping between the two parts, A_i and A_j, are the areas of the two parts. This feature indicates the similarity between the two areas.

2. Normalized distance between y centroids:

$$D_y = \frac{|y_i - y_j|}{y_{max} - y_{min}}.$$

3. Normalized distance between x centroids:

$$D_x = \frac{|x_i - x_j|}{y_{max} - y_{min}}.$$

In this way we can establish the relative position of each part with respect to the position of the parts at the extremities (left, right, top, bottom) of the window.

4. Orientation difference between parts:

$$D_\phi = |\phi_i - \phi_j|,$$

where ϕ_i and ϕ_j are the angles of the respective Gabor filters of the overlapping parts.

7.4 Conditional Rule Generation System

Once the unary and binary features have been generated, they are passed on to an object recognition system known as Conditional Rule Generation (CRG) that generates a tree of hierarchically organized rules by taking into account label compatibilities between unary and binary features [18, 19].

The CRG system was designed based on the concept that objects and patterns can be described as being composed of a set of parts. The pattern descriptions involve unary (single part) and binary features. As a result, the set of features characterizes the structure of the pattern. In traditional pattern recognition systems, there is a so-called label compatibility problem. That is, two patterns with identical unary and binary features may be obtained from structurally different patterns; and patterns of different classes can share common feature spaces. This results in poor class discrimination due to similarity between intra-class and inter-class variations. The CRG system overcomes these difficulties by generating rules that satisfy the label compatibility constraints.

Figure 7.17 shows how the rules are resolved. First, the unary feature spaces for each part, $U = \{u(p_i), i = 1, \ldots, N\}$, where $u(p_i)$ are the unary features for each part p_i, are clustered into clusters U_i. Clusters which are unique with respect to class memberships have their rules resolved at this level (e.g., U_3 in Figure 7.17). Class membership is determined by calculating the cluster entropy,

$$H(i) = -\sum_j q_{ij} \log q_{ij},$$

where q_{ij} is the probability of elements in class i belonging to class j. Each of the unresolved clusters is further analyzed using binary features by constructing a binary feature space $UB_i = \{b(p_r, p_s) | u(p_r) \in U_i \text{ and } S(p_r) = S(p_s)\}$, where $S(p_k)$ is the sample to which p_k belongs. This feature space is clustered with respect to the binary features into clusters UB_{ij}. Again, clusters which can be resolved with respect to the class membership at this level provide the classification rules, otherwise UB_{ij} is further analyzed with respect to the unary features of the second part and the resulting feature space $UBU_{ij} = \{u(p_s) | b(p_r, p_s) \in UB_{ij}\}$ is clustered into clusters UBU_{ijk}. The above procedure repeats until all clusters have been resolved or a maximum level has been reached. If a cluster has not been resolved at the maximum level of the tree, the clusters are broken into smaller subclusters making them easier to discriminate.

As the CRG system uses non-incremental learning, the feature spaces for all of the training words are first generated, combined, then passed to the training module. This produces the rules used by the recognition module. The feature space for the test word is created, as outlined above, and passed to the recognition module along with the generated rules. Each part is then classified as belonging to the most likely word class. The word class with the highest number of matching parts is then considered the recognized word.

7.5 Experimental Results

To test this system, we extracted a word database from the same database (CEDAR CDROM1) that has been used in the previous chapters. These words consisted of USA city names which have a sufficient number of samples and are similar in size and structure. Both printed and cursive writing samples were used. We used a total of 12 words for the experiments with 238 samples being selected - 102 randomly chosen samples for training, and the remaining 136 samples for testing. Figure 7.18 shows the recognition rates for four of the individual words and the overall recognition rate of the 12 words, and shows the recognition rate for the words being selected in the n top classification positions.

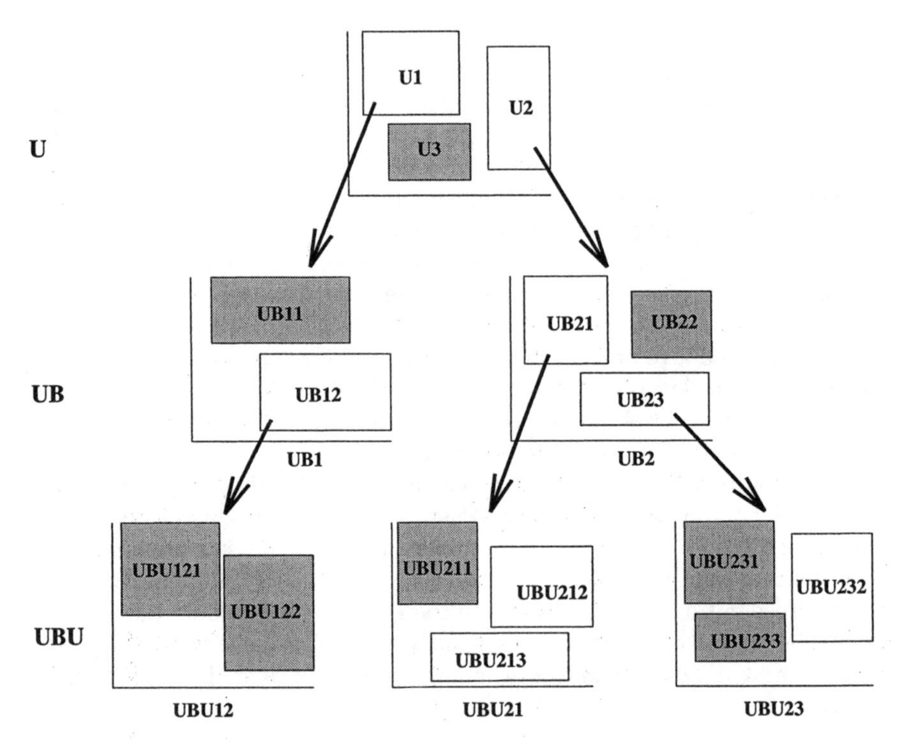

FIGURE 7.17. The Conditional Rule Generation procedure. Here the unresolved unary clusters (U_1 and U_2) - having more than one class represented in each cluster - are expanded to binary clustering of UB_1 and UB_2 feature spaces. This process is continued until either all rules are resolved or the predetermined maximum rule length is reached. The shaded blocks indicate the resolved clusters [18].

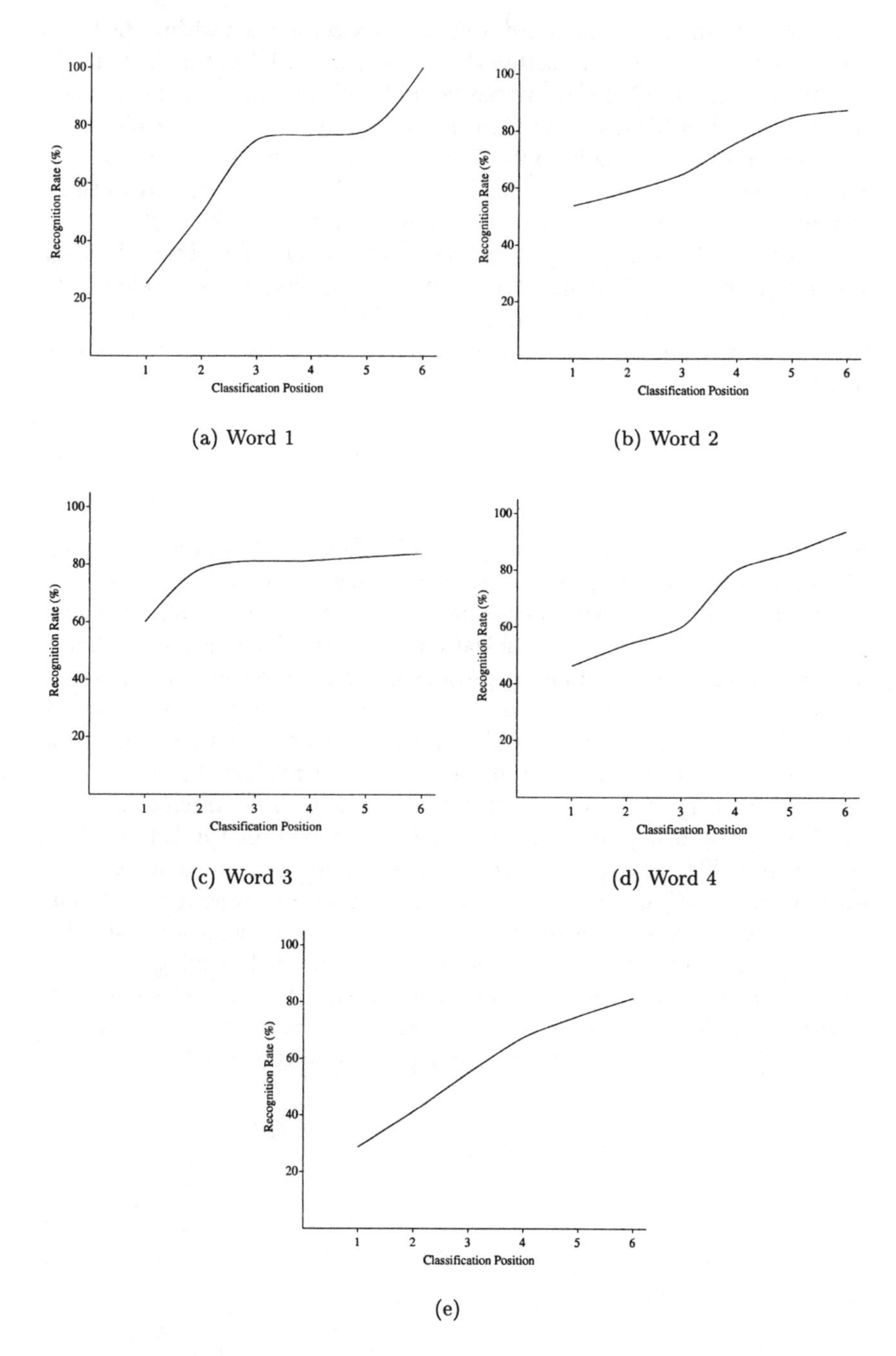

(a) Word 1

(b) Word 2

(c) Word 3

(d) Word 4

(e)

FIGURE 7.18. Percentage recognition rates for the word being selected in the n top classifications, (a-d) sample words, (e) overall.

The results show that the word was correctly recognized within the top 3 choices 54% of the time and within the top 6 choices 81% of the time. These results are comparable to other results published recently. The recognition rate for the word being classified in the top 5 selections, Ho *et al* achieved recognition rates ranging from 92.2% to 96.1%, and Paquet and Lecourtier achieved recognition rates ranging from 64% to 78.5%, though both of these results were with much larger training and testing sets [109, 192].

At present, there is only, on average, 8.5 samples per word class for training. Even with the current 8.5 samples per class, we were able to to obtain approximately 61% recognition rate. It is envisaged that with more training data using the structural and relational approach, the recognition rate will be improved.

7.6 Conclusion

In this chapter, we have presented a word recognition system capable of recognizing unconstrained handwritten words. In this system, features are extracted directly from gray-scale images without performing any preprocessing (binarization, skeletonization). This is advantageous in that it does not introduce distortion or spurious artifacts into the word image.

We have also presented two new approaches to determining the global slant of a word image. We consider the Fourier domain approach as the better of the two, because it produces more consistent results.

The recognition is based on oriented parts that are extracted from the word image by using Gabor filters. The parameters for the Gabor filters are calculated from each of the word images, thus the part extraction from each word is optimal. From the extracted parts, we generate unary and binary features for classification. The recognition is implemented using the CRG system. Being a *part-driven* system, instead of classifying the whole word as one unit, the CRG classifies each part of the word as belonging to a word class. It recognizes words by selecting the word class that contains the maximum of part classifications for a given written word.

8

Handwritten Word Recognition Using Fuzzy Logic

8.1 Introduction

As we discussed in Chapter 7, words (or characters) are entities composed of parts or segments at different orientations, lengths, and positions. We can recognize handwritten words by parts using unary and binary features. In this chapter, we introduce a new holistic approach to handwriting recognition that uses fuzzy logic for handling uncertainties in written words and measuring similarities.

Although handwritten words (or for that matter, characters) vary significantly in styles by different writers, the *structure* of the word remains consistent over all writers. From cognition point of view, it is this consistency that provides *normative* clues to enable the human reader to recognize the word with high accuracy. For instance, a *t* consists of a tall vertical stroke and a crossing horizontal stroke; an *o* is more or less a circular shape.

Usually, a significant amount of writer variation can be removed by preprocessing; and we can use slope and tilt correction to make the handwritten word appear in an up-down fashion along a horizontal line, as with the printed words on this page. However, due to differences in writing styles, tools, media, etc, we are still left with incomplete and very often ambiguous information that makes handwriting recognition difficult. In this chapter, we present a fuzzy model to represent both the sizes and positions of the extracted word features. As in Chapter 7 we use Gabor filters to extract oriented word parts. Using structural features and fuzzy models, we measure the degree of match between the word features and that of the training words. We show that fuzzy models can effectively handle variations. For this method, we recognize the word based on the closest match by a fuzzy decision process.

8.2 Extraction of Oriented Parts

We will use the process introduced in Chapter 7 by a bank of Gabor filters for extracting oriented features from gray-scale word images. To do this we first correct the slant and tilt of the input word to remove variations between different writing styles. The result is a normalized image with the word upright and level, relatively independent of writing styles.

8.2.1 Slant and Tilt Correction

The slant and tilt-corrected words produce a consistent set of features over a range of input words, enabling easier combination of the word features in the training phase. We use the frequency-domain slant correction methods introduced in Section 7.3.1. The slant angle is illustrated in Figure 7.11 using a gray-scale image of the word 'Silver'.

The *tilt* of the word is defined as the upward or downward trend of the writing with respect to the horizontal base line. The tilt angle is shown in Figure 8.2 for a gray-scale image of the word 'Falls'. Figure 8.2(a) is the tilt angle for Figure 8.2(b). Figure 8.2(c) depicts the word after tilt correction.

In this section we introduce a simple, yet effective method for tilt correction. In this method we need to find the baseline of the word, taking the descenders into consideration. This is accomplished by finding the maximum difference in the positive gradient of the distribution of the horizontally projected histogram [227], i.e., the sharpest slope as shown in Figures 8.1. This process can be described by

$$\text{tilt} = \text{maxarg}\left(\max_{j\in[s-1,N/2]}\left(\sum_{k=0}^{s} h_\tau(j+k+1) - \sum_{k=0}^{s} h_\tau(j-k)\right)\right) \quad (8.1)$$

where s is the size of the sliding window, N is number of rows of the image, $h_\tau(i)$ is the histogram formed through skewing the image using $x' = x, y' = y - x\tan(\tau)$ for each of the angles in the range $R = \pm 20°$. The size of the sliding window needs to be large enough so that small variations are not misinterpreted, but small enough not to blur the result. A good compromise used is $s = 5$. Figure 8.2(c) shows the tilt correction applied to the word 'Falls' in Figure 8.2(b.)

8.3 System Training

After word features have been extracted, we need to train the word recognition system with a set of handwritten word samples. The training phase forms a fuzzy representation of each word group W_j. A *word group* is a set of samples of the same word. These samples are derived from different written versions of the single word. For example, the two samples for the word 'Baton' shown in Figures 8.3(a) and (b). The fuzzy representation is used to classify an unknown word as one of the training words.

Formally, let $\{W_1, W_2, \ldots, W_N\}$ be N different words. For example, we may define W_1 as the word group for 'Baton' which may contain N_j different written versions, i.e., $W_1 = \{W_{11}, W_{21}, ..., W_{N_j,1}\}$. In general, let $\{W_{jk} | k = 1, 2, \ldots, N_j\}$ be the N_j training samples for word W_j, $1 \leq j \leq N$. Without loss of clarity, we may simply call the set $\{W_{jk}\}$ the training word-group for word W_j. The word W_j is in terms of its structural features which are a

FIGURE 8.1. Horizontal projection histograms: (a) the slope-corrected word 'Vegas', (b) its horizontal projection histogram, (c) the tilt-corrected image and (d) its horizontal projection histogram.

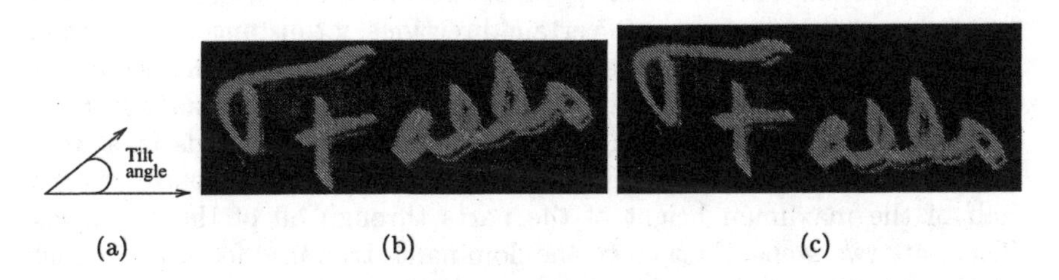

FIGURE 8.2. Results of tilt correction for the maximum difference of distribution method: (a) the tilt angle, (b) slope corrects words, (c) tilt correction applied.

set of n oriented-feature images generated by the bank of Gabor filters as those in Chapter 7. To form a fuzzy representation of the word W_j we first find distribution areas of the major structural features of the word, then fit a set of 2-D fuzzy membership functions to these distribution areas.

For consistency and clarity, we use the following subscript notations throughout the remainder of the chapter:

- i refers to an orientation index, for instance, a Gabor orientation angle ϕ, $i = 1, 2, \ldots, n$, where n is the number of orientations used ($n = \frac{180}{\Delta\phi}$, $\Delta\phi = 15°$, $n = 12$).
- j refers to the word group index, $j = 1, 2, \ldots, N$, where N is the number of word groups and also the lexicon size for a problem domain.
- k refers to a particular training sample of W_j, $k = 1, 2, \ldots, N_j$, where N_j is the number of training samples used for word W_j.
- l refers to one of the membership functions μ_{ijl} resulting from the training process for word W_j and orientation i, $l = 1, 2, \ldots, N_{ij}$, N_{ij} is the number of membership functions for W_j.
- (x, y) refers to a particular point (pixel) in the image, $x = 1, 2, \ldots, N_x$, $y = 1, 2, \ldots, N_y$, where N_x, N_y are the width and height of the normalized images, respectively.

The following sections present each of the stages used in forming the fuzzy representation of the word. The fuzzy models are also defined.

8.3.1 *Word Alignment*

Once the structural features for each training word in a word group have been extracted, the next phase in creating the fuzzy models is to combine the structural features for each of the oriented views to form a composite standard feature image for the word group. We need to determine the *dominant* features to weed out spurious information. In general, dominant features are typically strokes in certain directions; for instance, in the letter 'H', the dominant features are oriented mostly vertically, whereas in the letter 'E', the dominant features are horizontal. These dominant features are normally consistent throughout the set of training words in a word group. In this chapter, we define the dominant features that are at least half of the maximum height of the parts through all of the 12 angles. There are two steps: First, align the dominant structural features for each of the words; and the second step, mark the position of the dominant features for each feature orientation and transform the feature images into a standardized form.

Alignment of the dominant features in the word is necessary to take into account of different writing styles of the samples and of the formation of a model of the word group, which is obtained by combining the training samples. Figures 8.3(a) and 8.3(b) show an example of the word 'Baton' written by two different writers. The corresponding features (at $\phi = 90°$) are shown in Figures 8.3(c) and 8.3(d). The figures show, for example, that the relative horizontal position of the vertical features of the letters **B** and **t** in Figures 8.3(c) and 8.3(d) do not align properly. By aligning each of the training words in a word group to its dominant features, we are able to generate a consistent set of features for the word group.

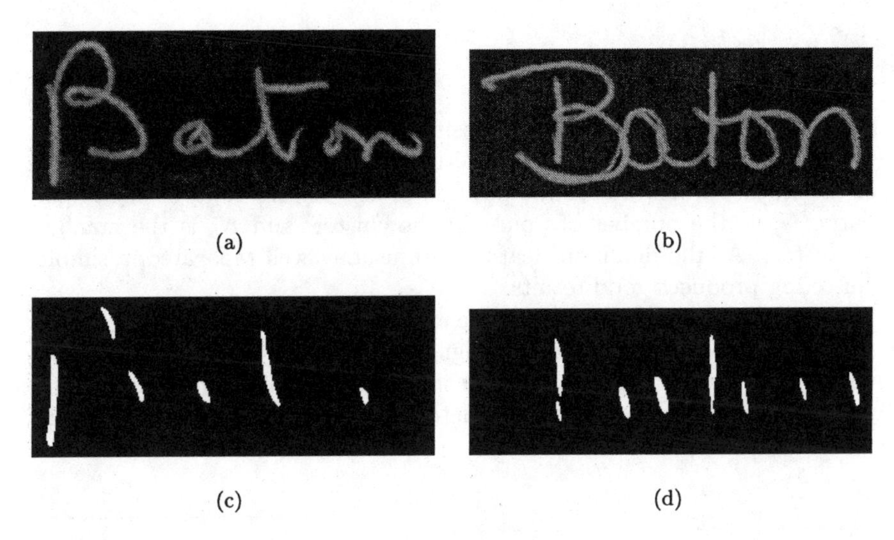

FIGURE 8.3. Example of two slant and tilt-corrected words written by different writers. (a), (b) original images; (c), (d) extracted features for $\phi = 90°$.

As shown in Figure 8.3, the features of 'Baton' at $\phi = 90°$ are vertical segments. Although due to different writing styles there are some size and location variations, the distribution of the segments for Figure 8.3(a) is similar to that for (b). We can align the word features both horizontally and vertically by finding a common set of relative alignment points within the features. The horizontal alignment points are found by locating the positions for this word, the dominant features for word group W_j. The dominant features are located in the angle range $[30°, 150°]$ that covers variations in slant correction and character shapes.

Figure 8.4 depicts the alignment process. First we compute the centroids of the oriented feature segments in each of the images in the word group W_j. These centroids are then clustered using a standard hard k-means clustering algorithm [17, 242]. The number of clusters is determined from the dominant features. The value of the centroids are normalized to the size of the image, making their position independent of the size of the image. The location of the cluster center for a feature is calculated by taking a weighted average of the cluster centroid points found for this feature in the N_j samples of word group W_j. The weighting factor is defined as the area (number of pixels) in the feature segment. This biases the value towards the location of the larger segments. We find the horizontal alignment points by the following formula:

$$C_p = \frac{1}{N_p} \frac{\sum_{b=1}^{N_p} a_b P_b}{\sum_{b=1}^{N_c} a_b}, \tag{8.2}$$

where C_p is the dominant feature position, $p = 1, 2, \ldots, N_c$, P_b is the horizontal coordinate of the centroid position within the cluster, a_b is the area of the feature which biases the position of the feature towards the larger parts, N_p is the number of points in the cluster, and N_c is the number of clusters. As the dominant features are usually well separated, a simple clustering produces good results.

We determine the vertical alignment points simply from the baseline of the word image. In this process, we ignore descenders (e.g., the descending tail of g) as they do not contribute to the position of the word. This is done using a method similar to that for determining the tilt of the word image (see Section 8.2.1).

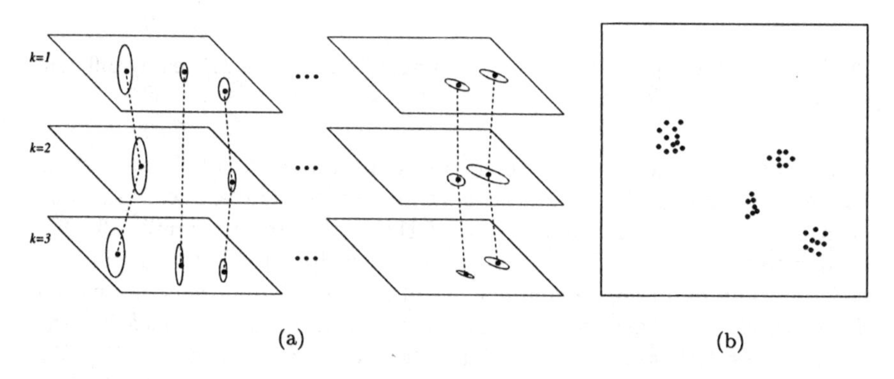

FIGURE 8.4. Alignment of the oriented features: (a) Similar sizes and positions of the features' centroids are used for matching, where the dots represent the centroid of each feature, ϕ_i indicate orientation of the features. (b) An illustration of the clusters for the samples in a word group with features oriented at, say, ϕ_i.

After we have found the horizontal and vertical alignment points (the dominant feature positions), we transform the extracted feature images into a standard data structure of aligned image features. In the standard data structure, we normalize image intensity values to the range $[0, 1]$ and consistent image size (N_x, N_y = the size of the smallest image dimension in the set of images of the word group W_j). With the normalized locations of the dominant features, we can re-align each of the images for the word group W_j to the position of the dominant features, and re-size the image to the standard size. Figure 8.5 illustrates the process for horizontal alignment. In the recognition process, the alignment points for each word group are stored and later used to align input words.

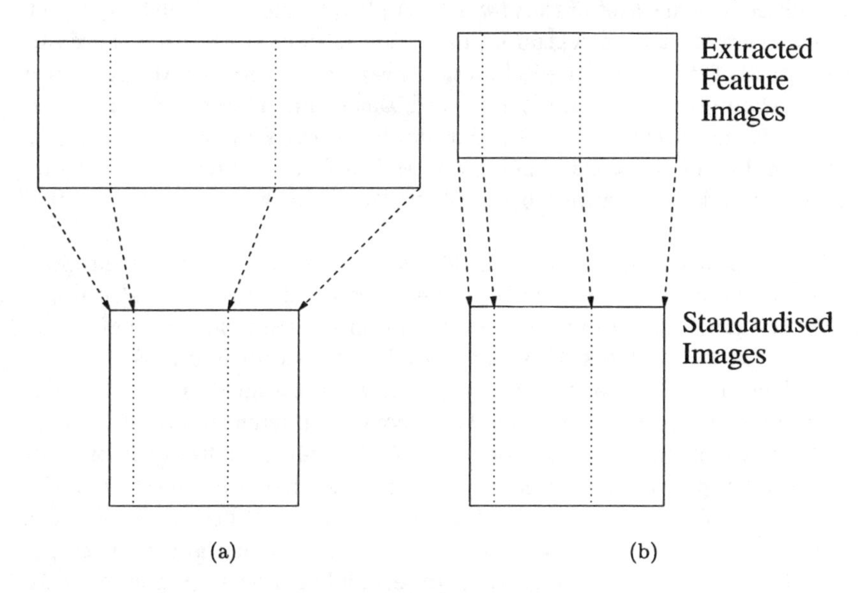

FIGURE 8.5. Alignment and transformation of two differently sized features images (a), (b). The features images are shown on the top, and the standardized images are shown on the bottom. The dashed lines represent the relationship to the horizontal relative alignment points.

Because the alignment points are relative to the size of the image, the *structure* of the word (relative to the position of the extracted features) will be maintained after the transformation. As each training sample of the word group W_j contains essentially the same feature structure, we can obtain the major structural components by simply adding the standardized images together. The value at (x, y) of the composite image R_{ij}, $1 \leq i \leq n$, $1 \leq j \leq N$ is:

$$R_{ij}(x,y) = \frac{1}{N_j}\sum_{k=1}^{N_j} W'_{ijk}(x,y), \qquad 1 \leq x \leq N_x\,, \quad 1 \leq y \leq N_y\,, \qquad (8.3)$$

where W'_{ijk} is the re-aligned image of W_{ijk} and N_j is the number of training samples in word group W_j. Figure 8.6(g) illustrates this idea: It is the composite image produced from the six images oriented at $phi = 90°$ (i.e.,

$i = 6$) as shown in Figures 8.6(a)-8.6(f), recalling that i is the orientation index.

8.3.2 2-D Fuzzy Membership Functions

This section describes how the fuzzy membership functions are generated from the training data. Several examples from each word group are selected at random. The features of these word samples are then analyzed to extract the structure of each word group. Finally, the extracted structure is used to form a set of fuzzy membership functions, which characterizes the structure of word group W_j. For example, the word 'Baton' has a large vertical feature at the left end of the word ('B') and another vertical feature ('t') roughly at the middle of the word. These two vertical features are isolated by the white rectangles in Figure 8.6(h). We will explain why each one has two rectangles shortly.

To represent the spatial location of the dominant features, we propose fuzzy membership functions which have the general shape of *twisted trapezoids.* To define this peculiar shape, two oriented bounding rectangles (the top and bottom surfaces of the trapezoid as shown in Figure 8.6(h)) are used. The orientation of the upper and lower rectangles can be different because the two thresholded boundaries contain different information, e.g., the direction of the central (upper) part can be different from the broader outer (lower) part as the upper section contains a narrower selection of the overall data. We have introduced the twisted trapezoid because it provides a better fit to the shape of the data being modeled than regular 2-D trapezoids. Figure 8.7 depicts the twisted trapezoidal membership function idea, and Figure 8.9 shows an example.

The upper and lower boundaries of the membership function are determined from the upper and lower thresholding points of the aligned images $\{R_{ij}\, i = 1, 2, \cdot, n\}$. The lower threshold point also helps remove noise and insignificant features. The upper H_u and lower H_l thresholding points are determined by

$$H_{u_{ij}} = c_u \max R_{ij}(x, y), \qquad H_{l_{ij}} = c_l \max R_{ij}(x, y), \tag{8.4}$$

where $x = 1, 2, \ldots, N_x$, $y = 1, 2, \ldots, N_y$, x, y index the location of the point in the image R_{ij}, the constants c_u and c_l determine the thresholding points, and N_x and N_y define the size of the standardized image. In the current implementation, $c_u = 0.25$ and $c_l = 0.1$. The selection of these values allows for the general shape and size of the word group feature to be extracted while ignoring the background noise and insignificant features being used (thresholded out).

As a result of the alignment process, similar features from the words (e.g., the vertical strokes in letter 't' in Figure 8.6) are clustered together across different writing styles. An example of this combination of features is

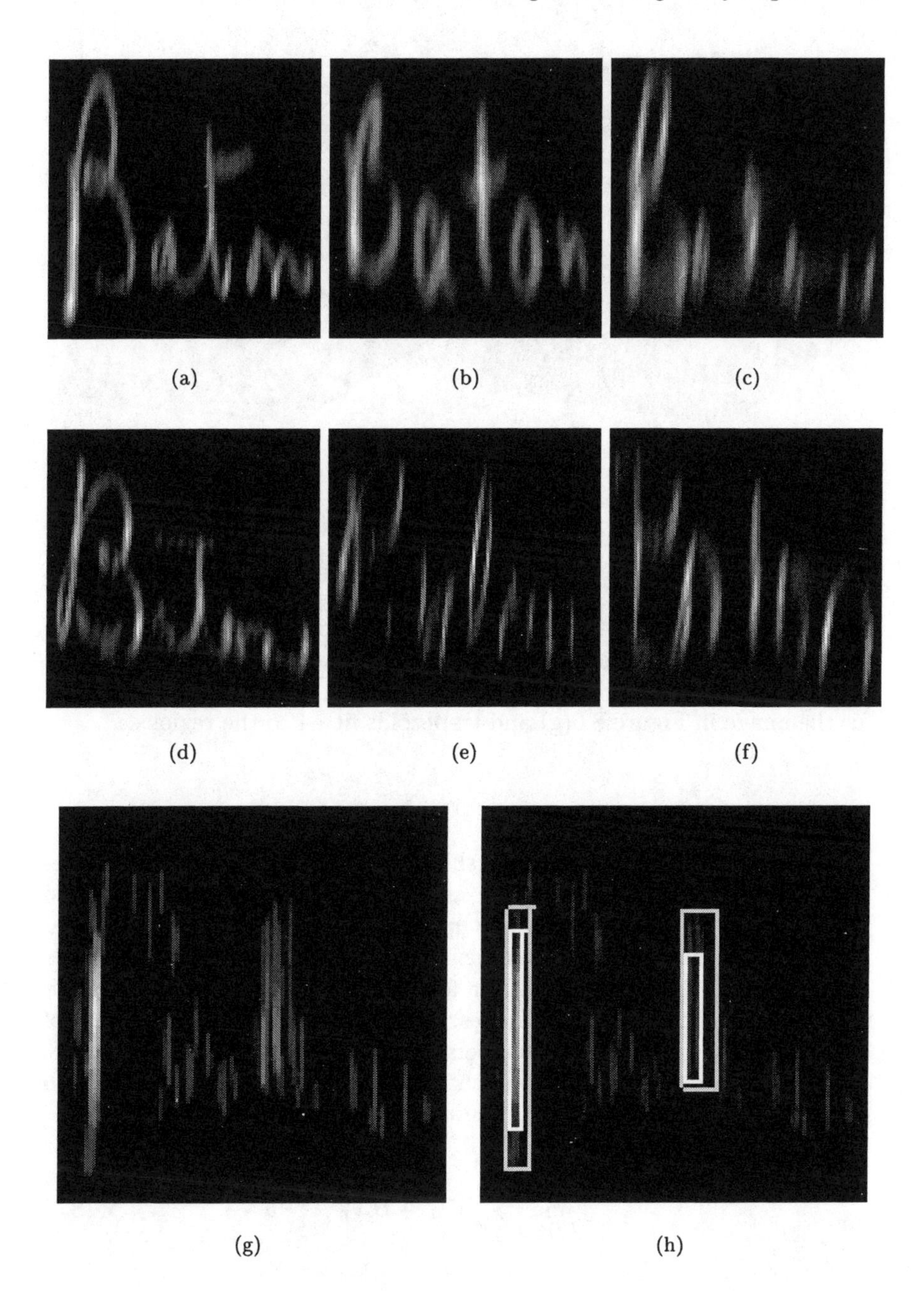

FIGURE 8.6. Fitting of membership function to word 'Baton'. The images shown are for Gabor filter angle $\phi = 90°$, (i.e., $i = 6$). (a)-(f) Re-aligned and aligned images of the samples that are *combined* to form the composite image in (g), (g) composite aligned image R_{6j}, (h) membership functions fitted to dominant areas in (g).

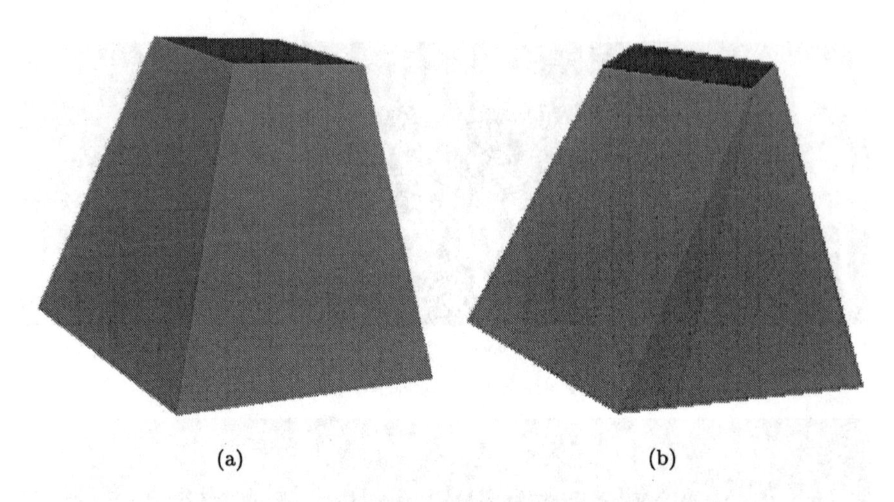

(a) (b)

FIGURE 8.7. (a) Trapezoid, (b) Twisted trapezoid.

demonstrated in Figure 8.6. Figures 8.6(a)-(f) show the normalized images of six examples of the word 'Baton', and Figure 8.6(g) shows the composite of these images. Figure 8.6(h) shows an example of the features clustered for the image in Figure 8.6(g) and trapezoids fitted to the regions.

Corner Point Correspondences

To correctly construct the membership functions, the corner point correspondence between the top (inner) and bottom (outer) rectangles has to be found. This happens because the upper and lower rectangles are created independently and can be of different orientations and sizes. The corner correspondence also depends on the amount and sense of the relative orientation between the two rectangles. There are two practical schemes of constructing the sides of the membership function as shown in Figure 8.8. We propose a connection scheme that results in the shortest connection distances between the corners of the two rectangles:

$$\min_{(i,j)\in C} \sum_{k=1}^{4} \|T_{ik} - B_{jk}\| , \tag{8.5}$$

where C is the set of valid combination pairs (i, j) connecting a top rectangle corner to a bottom rectangle corner (24 possible combinations), T represents the coordinates of a top rectangle point, B represents the coordinates of a bottom rectangle point, and $\|\cdot\|$ is the distance (the Euclidean norm) between the two corners.

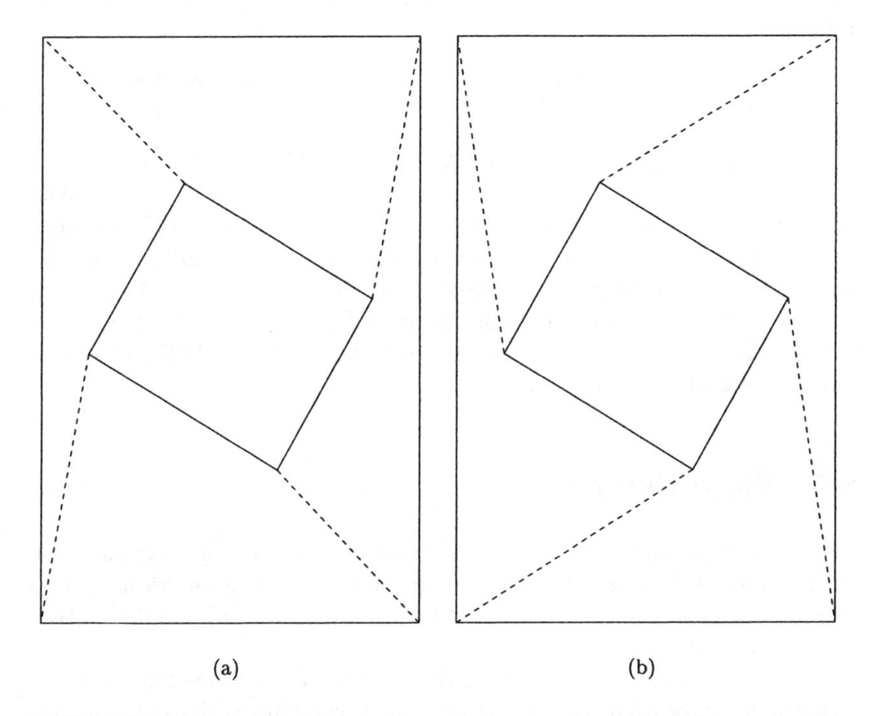

FIGURE 8.8. Two possible schemes for constructing the sides of the membership function. Note, for illustration purposes we have exaggerated the orientational differences between the top and bottom rectangles, which in most applications are small.

Membership Value

Figure 8.9 shows a twisted trapezoidal membership function in 3D. The membership value $\mu(x, y)$ of each point in the lower rectangle is determined by one of the twelve equations, three of which are given in Eq.(8.6), based on the four region types shown in Figure 8.9(a). To calculate the height in the sloped areas, we further divide the area into regions, with each of the four faces being divided into three subregions. Subdivision of one of the four faces is shown in Figure 8.9(a). Calculation of the membership value for any (x, y) in the union of regions $1-4$ in Figure 8.9(a) is done as follows:

$$\mu(x,y) = \begin{cases} 1.0 & \text{if } (x,y) \in \text{area 4;} \\ \frac{x - x_{B_a}}{x_{T_a} - x_{B_a}} & \text{if } (x,y) \in \text{areas 1 and 3;} \\ \frac{(x - x_{B_a})(y_{T_b} - y_{T_a})}{(y - y_{T_a})(x_{T_b} - x_{T_a}) + (x_{T_a} - x_{B_a})(y_{T_b} - y_{T_a})} & \text{if } (x,y) \in \text{area 2.} \end{cases} \tag{8.6}$$

where $\mu(x, y)$ is the membership value at point (x, y), B and T represent the bottom and top rectangular features of the membership function, B_a represents the bottom left hand corner (as shown in Figure 8.9(a)) and, x_{B_a} represents the x coordinate of the point B_a. The other three sides use a set of equations similar to that shown in (8.6). Figure 8.9(b) shows an example of a membership function.

8.4 Word Recognition

This section presents the word classification procedure and explains the methodology behind it. The first stage of the word recognition process extracts the oriented features from the word image in a way similar to that for the training words.

The set of oriented features obtained from the Gabor filters are processed to extract the locations of the dominant features. This is the same process as that described in Section 8.3.1 for the training words; except it is being performed on a single word as opposed to a set of words (word group) as done during training. The set of feature images $\{V_i\}$ for the input word is then aligned, size and intensity normalized according to each of the alignment points $\{P_i\}$ found during the training phase for each of the words W_j. The alignment process aligns the dominant features of the test image to each of the dominant features of the trained word. Since there are usually only one or two dominant features, even if the number of features are different, for similar words their positions will still be similar. For dissimilar words, the position of the dominant features are likely to be different, resulting in the word being poorly aligned, producing lower classification results. The alignment process is also required for the test word to ensure that the positions of its features align with the position of the dominant

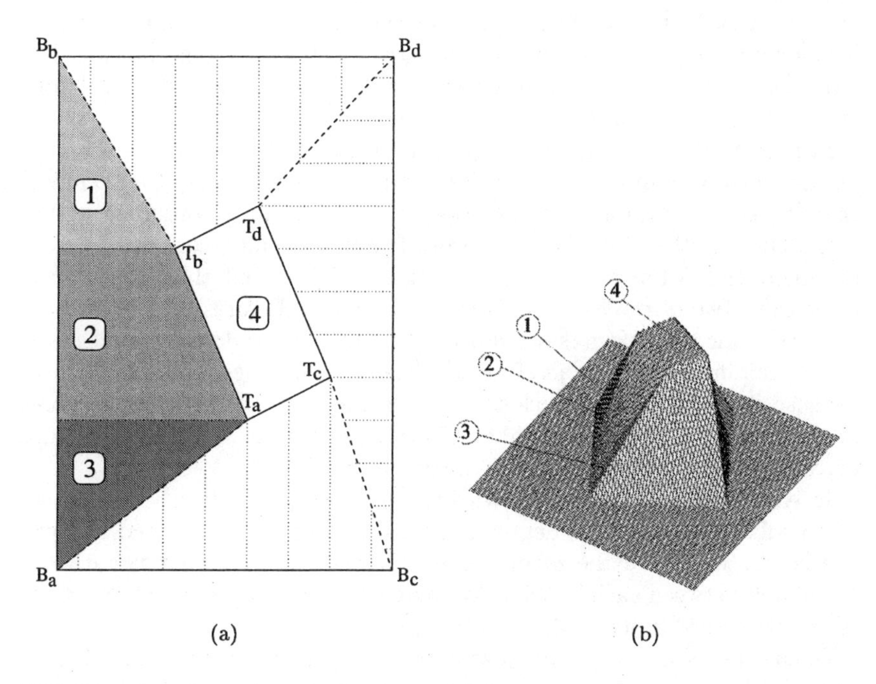

FIGURE 8.9. 2-D membership functions: (a) Structure and areas of a membership function (a top view), (b) Example of a single membership function (a perspective view).

features of the trained words as that shown in Figure 8.3. This set of aligned images $\{V_i^*\}$ is then compared using a fuzzy decision process to each of the training word groups using each of the membership functions of the words W_j. And the closest match is selected. The following sections describe the matching process in detail.

8.4.1 Fuzzy Decision Making Process

To illustrate the fuzzy decision making process, we consider the scenarios in Figure 8.10 which depict the extracted, normalized features of a test image (shaded areas) at $\phi = 90°$, superimposed on the membership functions for four different training word groups (that is, four different words). The objective is to select the training word that best matches the test word (essentially, nearest prototype classification).

Inspecting the four examples shown in Figure 8.10, we can see that the extracted features of the test word seem to correspond best to the membership functions shown in Figure 8.10(a), but are not fully contained within them. On the other hand, the test word features are fully contained within the membership functions shown in Figure 8.10(b), but they do not correspond to two of the four membership functions. Figures 8.10(c) and (d) show that the test features are not fully contained nor do they correspond to the training features. Therefore, the (visual) choice here is Figure 8.10(a) because it seems to be the closest match between the extracted features of the test word and the membership functions both in terms of feature correspondence and containment (degree of match).

In each of the four cases shown in Figure 8.10, a similarity rating can be calculated showing the degree of match between the extracted features of the test word and the membership functions. This similarity rating is calculated between each of the n extracted feature images, and each of the N training word groups $\{W_j\}$, as follows.

Define the $(n \times N)$ *similarity matrix* S to be the matrix arranged as:

$$S_{ij}(W) = S(W; \{\phi_i\}, \{W_j\}) = \begin{array}{c|cccc} & W_1 & W_2 & \cdots & W_N \\ \hline \phi_1 & S_{1,1} & S_{1,2} & \cdots & S_{1,N} \\ \phi_2 & S_{2,1} & S_{2,2} & \cdots & S_{2,N} \\ \vdots & \vdots & \vdots & & \vdots \\ \phi_n & S_{n,1} & S_{n,2} & \cdots & S_{n,N} \end{array} , \quad (8.7)$$

where W is the input word, $\{W_j\}$ are the training word groups, $\{\phi_i\}$ are angles (of the Gabor filters) at which the features have been extracted, and S_{ij} is a fuzzy similarity rating between the test word and the membership functions of the training words for $i = 1, \ldots, n$, $j = 1, \ldots, N$. Next we describe how the similarity rating S_{ij} is calculated.

Figure 8.11 shows three trapezoidal membership functions, which are an illustrative slice through the membership functions for a word W_{ij}. Some

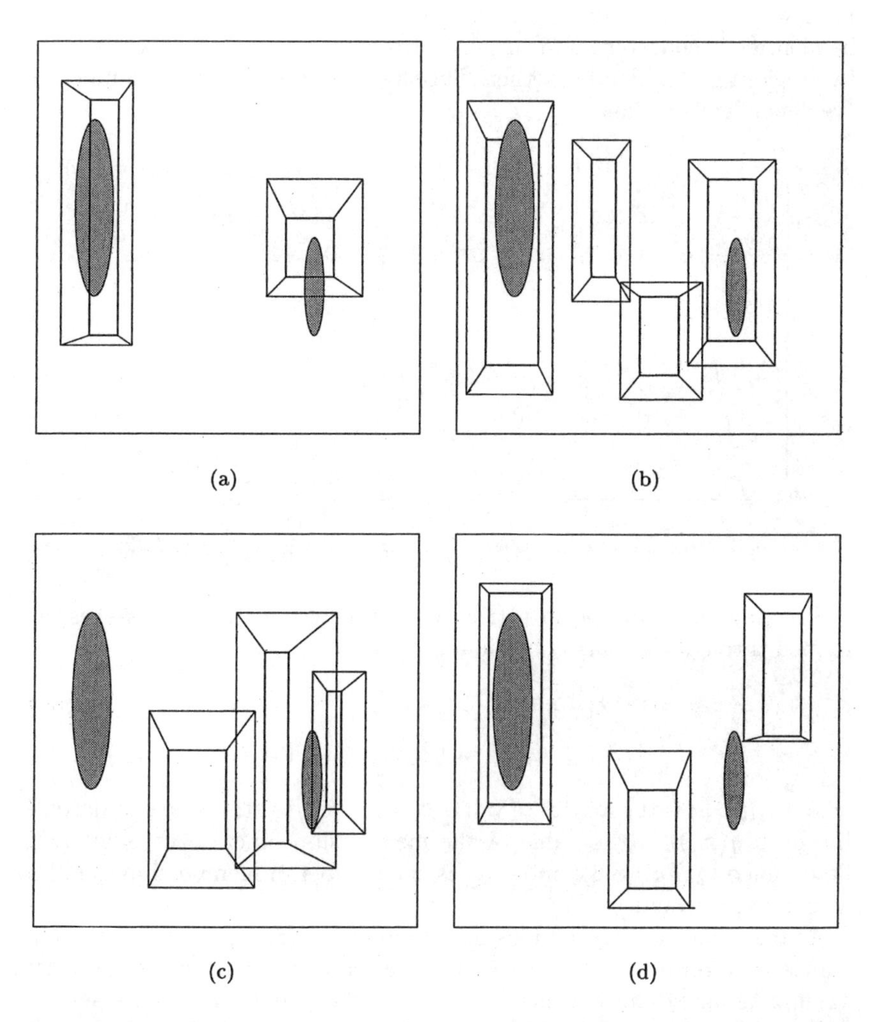

FIGURE 8.10. Matching a test word (shaded areas) to the membership functions obtained from four training word groups. The shaded areas are the extracted features (at $\phi = 90°$) of a test word and the superimposed rectangles represent the membership functions of the prototype words.

trapezoids such as A and B in Figure 8.11, overlap because of the fitting process. Thus, an oriented feature (e.g., a vertical stroke in letter 't' at x in Figure 8.11) is likely to be in a region with a high membership value $m_A(x)$, but can also have a smaller membership value $m_B(x)$ in a different region. For a word that has a double 't', the two likely regions of the 't' strokes will be identified separately, and the extended areas are likely to overlap as the two regions could blend together. The shapes of the membership functions are dependent on the data.

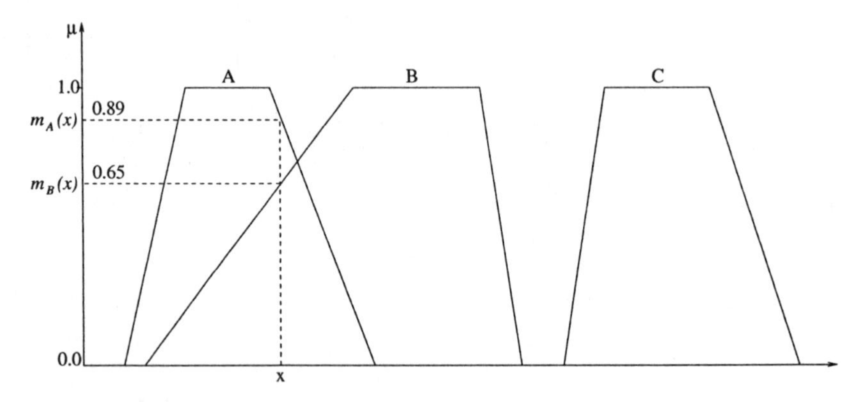

FIGURE 8.11. A slice through a set of membership functions.

Selecting the value of an attribute (the defuzzification process) is performed through the *max* operation,

$$a_{ij}(x,y) = \max_{l}\{\mu_{ijl}(x,y)\}, \qquad (8.8)$$
$$i = 1,2,\ldots,n; \quad j = 1,2,\ldots,N; \quad l = 1,2,\ldots,N_{mf},$$

where $a_{ij}(x,y)$ is the attribute value at point (x,y) within the standardized image; $\mu_{ijl}(x,y)$ is the value of the membership function at point (x,y) determined by Eq. (8.6); and N_{mf} is the number of membership functions associated with image W_{ij}.

As the membership functions may overlap, Eq.(8.8) selects the attribute values that represents the best of the alternatives (the most significant value). As an example, using Figure 8.11, the point x may be associated with membership function A (the vertical stroke of the letter 'K' in word W_{ij}) having a membership value: $\mu_A = 0.89$, and with membership function B (the vertical stroke of the letter 't' in word W_{ij}) having a membership value: $\mu_B = 0.65$. Thus the point fits into the region that has the maximum of the trained alternatives. The disadvantage of taking only the *max* value is that the influence of adjacent regions may be ignored, which may result in errors in some cases.

Let V be the test word (when features are generated, there are n images on ϕ_i). Next we need a rating of how well the features of the test word V_i

match the corresponding features of the training word W_{ij}. This similarity rating S_{ij} is calculated using a weighted average,

$$S_{ij} = \frac{\sum_{x,y} w_{ij}(x,y)\, a_{ij}(x,y)\, V_i'(x,y)}{\sum_{x,y} w_{ij}(x,y)}, \tag{8.9}$$

where $w_{ij}(x,y)$ is the weight at (x,y) in the training word W_{ij}, $V_i'(x,y)$ is the value of the same pixel (x,y) in the input word V_i normalized to 1. The weights $\{w_{ij}(x,y)\} \subset [1,0]$ indicate the relative importance of each attribute value. If $w_{ij}(x,y) = 0$, it indicates that the attribute is of no importance, whereas a weight of 1 indicates maximum importance.

The weight $w_{ij}(x,y)$ is calculated from

$$w_{ij}(x,y) = \begin{cases} w'_{ij}(x,y) \text{ if } q_{ij}(x,y) \in \{W_{ijl}(x,y)\} \\ q_{ij}(x,y) \text{ if } q_{ij}(x,y) \notin \{W_{ijl}(x,y)\} \end{cases}, \tag{8.10}$$

where $w'_{ij}(x,y)$ is defined significance weighting in (8.11), $q_{ij}(x,y)$ is the point in the standardized image W'_{ij} of the test word (the location of this point corresponds to the location of the point (x,y) in training word W'_{ij}), $\{W_{ijl}(x,y)\}$ is the set of points where $\mu_{ijl}(x,y) > 0$, and $\mu_{ijl} = 0$, if $W_{ij}(x,y) \notin W_{ijl}(x,y)$. The condition, $w_{ij}(x,y) = q_{ij}(x,y)$, is included to cater for the points in the test image which do not lie within one of the membership function regions of the training words. This will penalize the S_{ij} rating as $a_{ij}(x,y)$ will equal 0.0 at that point.

A *significance of point* rating is used to determine the weighting factors $\{w'_{ij}(x,y)\}$ in (8.10). If a point appears in many membership functions of the training word groups, its contribution to determining the word group should be minimum, as it provides little discriminating information between word groups. On the other hand, if a point appears in the membership function of only one word group, its significance in distinguishing among the word groups should be maximum. Based on this observation, we define the following weighting function:

$$w'_{ij}(x,y) = \begin{cases} \frac{N-N_+}{N-1}\frac{1}{N_+}\sum_{j=1}^{N} a_{ij}(x,y) \text{ if } N_+ > 0, \\ 0 \qquad\qquad\qquad \text{if } N_+ = 0, \end{cases} \tag{8.11}$$

where N is the number of training word groups, and N_+ is the number of word groups in which $a_{ij}(x,y) > 0$ for a feature point of the test word. The effect of N_+ in (8.11) is illustrated in Figure 8.12. If $N_+ = 1$, the range of values for $w'_{ij}(x,y)$ is [0, 1]. If $N_+ = 2$ (the situation depicted in the shaded region of Figure 8.12), the range of values for $w'_{ij}(x,y)$ is [0, 0.75]. Thus, the value of N_+ scales the range of the significance ratings according to the number of word groups where a feature point of the test word is present. The more word groups in which a feature point appears, the smaller is the range of the significance rating for that feature point.

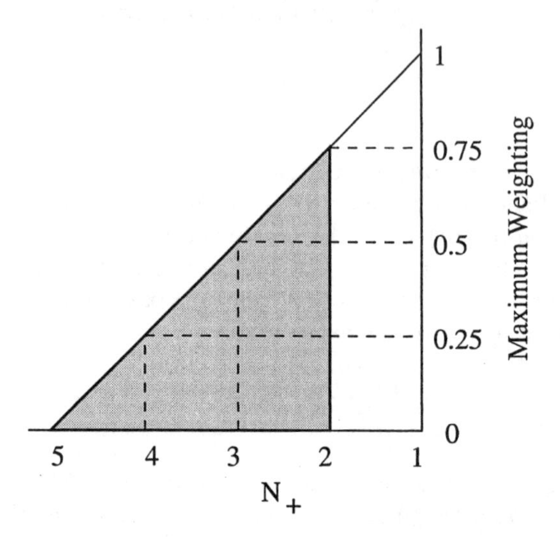

FIGURE 8.12. The range of values for the significance rating $w'_{ij}(x,y)$ in (8.11).

For example, if a feature point appears in membership functions of only one word group, then $w'_{ij}(x,y)$ may have a high significance rating, possibly 1, the maximum significance. In principle, the more word groups in which the feature point appears, the lower the significance rating of that feature point.

Selecting the word W_j that best matches the test word is done by aggregating the similarity measure values $\{S_{ij}\}$ using one of the several methods in [50]. In this chapter we use the *Simple Additive Weighted* method (SAW) [119]. The SAW method is chosen since it uses all n attribute values for each W_j, and each attribute has an associated importance weighting ν_{ij}. The importance weighting ν_{ij} is used to normalize each of the S_{ij} values that were derived from different data. The word classification label W^* can be determined with the SAW method as follows:

$$W^* = SAW(\{\phi_i\}, \{W_j\}) = \max_i \left\{ \frac{\sum_{j=1}^{N} \nu_{ij} S'_{ij}}{\sum_{j=1}^{N} \nu_{ij}} \right\}, \tag{8.12}$$

where

$$S'_{ij} = \frac{S_{ij}}{\max_i S_{ij}},$$

where S_{ij} is given by (8.9). The importance value is given by

$$\nu_{ij} = \frac{\Lambda_{mf_j}}{\Lambda_{T_j}} = \frac{\sum_{xy} a_{ij}(x,y)}{\sum_{ixy} a_{ij}(x,y)},$$

where Λ_{mf_j} is the total volume of the membership functions in word group component W_{ij}, Λ_{T_j} is the total volume of membership functions for word group W_j.

8.5 Experimental Results

To test our system, we take samples from the same word database used in the previous chapters which consists of USA city names with a sufficient number of samples from the CEDAR database [45]. Both printed and cursive writing samples were used. The 14 words shown in Table 8.1 (Number of Samples columns) were used for the experiments with 214 samples being selected - 102 samples were randomly chosen solely for training, and all samples were used for testing.

Table 8.1 shows the results of processing the test data after our system was trained with the training word groups. The table indicates the generalization percentages for correct labeling of each testing word. The values in the first column (for the Generalization Percentages columns) indicate the percentage that the word was classified as the highest rated word using (8.12). The second column indicates the percentage the word was classified as one within the top two rating words W_j from (8.12), and so on for columns 3, 4, and 5.

The last two words in the list, 'Terre', and 'Tulsa', were not classified with high ratings in the top position. 'Terre' was classified more as the word 'Sioux' and 'Tulsa' as 'Falls'. Though these words are dissimilar in appearance, there is a high correlation between the extracted oriented parts. For instance, at 90° there are high correlations between the features of the two words. The main problem with the proposed approach seems to be its granularity in blurring some of the details in the structure of the words. This was deliberately done to capture the dominant features of the written word. Therefore, there is a compromise between dealing with the huge variation in writing styles and textures, and capturing the fine texture in the features of the words.

At present, there are only, on average, 7.3 samples per word group for training. Even with this small set of training samples per group, we were able to obtain approximately 74% average recognition rate for the word being classified in the top position (column 1 in Table 8.1) and achieved 96.1% within the top five positions (average of column 5 in Table 8.1). These

		Number of Samples		Generalization Percentages				
j	Word W_j	Training Set	Testing Set	1	2	3	4	5
1	Baton	5	9	66.7	83.3	83.3	83.3	83.3
2	Boise	4	8	75.0	75.0	100.0	100.0	100.0
3	Dallas	9	19	77.8	88.9	100.0	100.0	100.0
4	Falls	15	32	85.7	92.9	100.0	100.0	100.0
5	Haute	4	7	100.0	100.0	100.0	100.0	100.0
6	Little	6	13	83.3	100.0	100.0	100.0	100.0
7	Louis	12	29	58.3	66.7	75.0	75.0	83.3
8	Moines	7	19	85.7	85.7	85.7	100.0	100.0
9	North	6	11	83.3	100.0	100.0	100.0	100.0
10	Salem	7	14	85.7	100.0	100.0	100.0	100.0
11	Sioux	12	24	91.7	91.7	91.7	100.0	100.0
12	South	5	9	60.0	100.0	100.0	100.0	100.0
13	Terre	4	8	25.0	75.0	100.0	100.0	100.0
14	Tulsa	6	12	33.3	50.0	66.7	66.7	83.3
			Average	**73.8**	**86.4**	**92.2**	**94.2**	**96.1**

TABLE 8.1. Word classification results. The five columns indicate the classification percentages of the word being placed within the top n positions. (Results are specified as percentages).

results are comparable to those reported recently in the literature. Other systems in the literature, for instance, considering the recognition rate for the word being classified in the top five selections, Ho *et al* achieved recognition rates ranging from 92.2% to 96.1% [109], and Paquet and Lecourtier achieved recognition rates ranging from 64% to 78.5% [192]. These results were based on much larger training sets, and used databases that are not readily available to the public. Therefore, the absolute recognition rates reported in the literature can serve only as an indication of our system's performance.

8.6 Conclusion

In this chapter, we have presented a word recognition system using fuzzy sets that classifies words with a similarity measure based on oriented word features.

We extract oriented word features from the word image using responses from a bank of Gabor filters. From the oriented feature images, we generate 2-D twisted trapezoidal fuzzy membership functions that characterize the structure of the word. We use a fuzzy decision process to classify unknown words.

These recognition rates are encouraging, especially in the light of the small number of training samples available from the database. It is possible to further improve the system's performance with different types of membership functions that provide more structural information and different fuzzy inference methods; for instance, Type-II fuzzy sets [133] that may enhance the discriminating power of the membership functions.

9

Conclusion

9.1 Summary and Discussions

This book has presented several approaches to recognizing handwritten numerals and words based on Markov models, conditional rules, and fuzzy logic.

We introduced several new methods for extracting features in Chapter 2. Generally, feature extraction is divided into two categories: extracting features from binarized images or from gray-scale images. The first step in both categories is pre-processing to reduce noise and variations in images. For handwriting recognition, the majority of systems extract features from binarized images, which require less memory and computation than that required by gray-scale images. Contours and skeletons are very popular in representing shapes of binarized images. Contour is an information lossless representation if the shape has only one closed contour. For simple patterns such as numerals and hand-tools, their shapes can be well represented by their outer contours [37]. The skeleton is a compressed form of shape representation and particularly suitable for representing patterns made up of line-segments such as handwritten images. Furthermore, skeletons can be easily obtained from binarized images by thinning algorithms that are available in the literature [150].

Binarization of gray-scale images, however, inevitably results in information loss. Extracting features directly from gray-scale images can reduce information loss. Wang *et al.* [251] and Lee *et al.* [156] proposed methods based on the first and second derivatives. The main disadvantages of these methods are that they are sensitive to noise and that the extracted skeletons may not be smooth. In Chapter 2, we presented a novel algorithm that is able to extract features from both binary and gray-scale images using Gabor filters. We also presented a method for estimating parameters of Gabor filters, so that Gabor filters can adapt the width of line-segments to produce the desired outputs. As a result, the proposed algorithm is robust to noise and insensitive to variations in line width and is able to produce better results than that of [251, 156].

HMMs have been widely applied in 1-D time series analysis, especially in speech recognition and econometric analysis [209, 101]. Recently, HMMs have been used for recognizing handwritten characters and words [190, 80]. In these systems, 1-D features were extracted from projected histograms or rows (columns) of 2-D images. However, experiments did not show good results compared to their counterparts for speech recognition. The motivation for using HMMs with contour-based shape representation for recognizing handwritten numerals lies in the following two factors: First, HMMs can be used to deal with large variations; and secondly, 1-D features can be easily obtained around contours. Although, HMMs are able to model local statistical information, they have difficulties to model structural information. To deal with this problem, we proposed a method using *macro-states* to model structural information. The method has been evaluated on the "bindigis" set and the "goodbs" set in CEDAR CDROM1 database and achieved 96.16% and 98.37% recognition rates respectively.

HMMs can be used in modeling statistics of spectral features. Because the magnitudes of contour-based Fourier spectra are invariant to rotation, translation, scale and reflection, contour-based Fourier descriptors are very popular methods in 2-D pattern recognition [178, 135, 198]. It is well known that there are considerable shape variations in handwritten digits. These variations are reflected in Fourier spectra. The reasons for using HMMs with spectral features to recognize handwritten numerals are threefold:

1. the states of HMMs can be used for modeling variations in spectra;
2. the state transition probabilities can be used for modeling constraints between spectral components including magnitudes and phases; and
3. the features can be readily arranged into sequences in the order of frequency.

The experimental results have shown that Cai and Liu's method can achieve 96.7% recognition rate on the "goodbs" set. In addition, this method can achieve high performance for recognizing hand tools [39].

Contours can well represent shapes of handwritten numerals and English characters; however, it is difficult for contours to represent patterns with complex structures such as Chinese characters, because contours of different Chinese characters often have significant correlations. Skeletons can be used to represent these patterns, therefore, it is desirable to develop handwriting recognition systems that use skeleton-based features as inputs. In Chapter 5, we presented a system using MRFs to model statistical and structural information. The experimental results of this system are comparable to most recently published results. Compared with the results in Chapters 3 and 4, the recognition rate of this system is slightly lower, due to the fact that the relaxation labeling algorithm may converge to a local optimum, whereas the Viterbi algorithm used in Chapters 3 and 4 can guarantee the global maximum. Furthermore, the MRF-based system is also effective in more general pattern recognition problems.

As discussed in Chapter 1, handwriting recognition approaches can be generally divided into two basic categories: segmentation-based and word-based approaches. Chapters 3, 4 and 5 have described segmentation-base approaches to handwriting recognition. In Chapter 6, we presented a holistic approach. MRFs provide a natural framework for modeling the spatial interactions among random variables in their neighborhood systems. Because MRFs enable us to recognize handwritten words based on the compatibility of the overall structural relationships between models and testing images, the system is robust to the order of features. Consequently, the recognition rates can be improved from 64.6% [38] to 69.0% for the first proposal and from 94.7% [38] to 96.5% among top five proposals.

In Chapters 7 and 8 we described two handwritten word recognition systems based on the holistic paradigm. In both systems, we extract word features directly from the gray-scale image without performing pre-processing such as binarization, skeletonization, etc. This scheme does not introduce distortion or spurious artifacts into the word image. Although some researchers have used the gray-scale image to generate skeletons [251], very few have considered the extraction of features directly from the gray-scale image itself. Word features are extracted from the word image using responses from a bank of Gabor filters. The parameters for the Gabor filters are calculated from each word image. In this way, the system is able to extract features *tuned* to the word's characteristics, resulting in a consistent set of parts for each word. In Chapter 7, from the extracted parts, we computed unary and binary features representing structural and relational features respectively, which are used in the recognition process using the conditional rule-generation (CRG) technique. Chapter 8 described a fuzzy logic approach to word recognition that classifies words with a similarity measure based on oriented word features. From the extracted parts, we generated 2-D twisted trapezoidal fuzzy membership functions that characterize the structure of the word. We used a fuzzy decision process to classify unknown words. The recognition rates are encouraging. It is likely that further improvements in recognition rate can be achieved if the system is trained with a large number of samples and uses new types of fuzzy membership functions and fuzzy inference logic.

9.2 Future Directions

Research and development on handwriting recognition have made some significant improvements, in particular, on-line handwriting recognition, For off-line systems, however, we are still unable to recognize handwriting with high reliability, accuracy, and robustness. In fact, we are far short of the performance level the human reader can achieve. In the following we present some major new approaches to further improving the handwriting system's performance.

Rejection

Until now, most methods indicate only the most likely candidate (digit or word). However, many applications require that the classifiers be able to give the best model for a given image and refuse to make decision if the explanation provided by the best model is unreasonably measured in terms of the output probability. For instance, 'M' is not a digit and the output probability of the best model will be very low (i.e., unreasonable), so that the classifier for recognizing numerals should be able to reject unlikely classifications. Sometimes, if the output probabilities of the best model and the second best model are very close, the classifiers should be able to reject the classification or leave it for further examination. For Markov model-based approaches, we can make a decision to reject a classification based on the output probability. Since the number of features (or observations) may differ from one image to another, the output probability can be normalized by the number of features. Therefore, we can reject classification if the normalized output probability is too low or the difference between the normalized output probabilities of the top-two models is too small.

Combining Classifiers

The motivation of combining classifiers is to utilize different representations and features by different classifiers to improve the reliability and accuracy of handwriting recognition. Generally, an individual classifier has its own advantages and disadvantages. The combination of classifiers with complementary properties is highly desirable. Many combination schemes have been proposed [143, 58, 256]. Experimental results have shown that in many cases the combination schemes outperform the best individual classifier. However, current combination schemes consider only the performance of classifiers over the whole database and ignore the information on particular patterns. For example, the overall performance of the method in [37] is better than Elms' method, but Elms' method can obtain better results in recognizing '0' [80]. Therefore, how to design the optimal combination of classifiers that can make full use of the available information is still an open problem.

Another interesting approach is to combine local and global model features and to extract features based on contexts and lexicons [162]. For instance, we may train the HMM with the usual local, segmented features with word-level features taking the advantages in both analytical and holistic approaches [6].

Reducing Computation for MRF Models

Currently, training and recognition in MRF-based approaches are the most time-consuming processes. In our systems, training and recognition are conducted by using relaxation labeling with the procedure of Highest Confidence First. It is well known that the speed of recognition is important to many applications. In order to speed up the recognition process, we limited the number of iterations to around 16 at the expense of the recognition accuracy. Further improvement would be to reduce the number of possible labels for the line-segment being processed. Some labels that are unlikely to be assigned to the line-segment should be removed from further consideration during the relaxation labeling process. In such a way, we may be able to reduce the computation. Further, it is interesting to investigate how to prune the possible label set for a line-segment without sacrificing recognition accuracy.

Multi-level Recognition Process

Instead of explicit segmentation, in this book we used the Gabor filter to extract oriented parts in handwritten words. This to a large degree avoids the uncertainty in segmenting subunits and the associated problems. To further improve the recognition, after pre-processing, we may extract oriented parts at different resolutions. For each training word, this generates a *feature hierarchy* that defines the word. At each level, we may model features using the MRF model to take into account both statistical and structural information. As a result, a word group can be represented by a feature hierarchy group. We may train the recognizer using the feature hierarchies. If at any level in the hierarchy, we have obtained the desired result, we can terminate the recognition process. Such an approach is expected to be computationally intensive. However, since computations at different levels are independent, it is possible to implement the recognition system in parallel.

Soft Computing in Handwriting Recognition

Since large variations and uncertainties present in handwritings, in most handwriting recognition techniques, researchers have used deterministic and statistic methods for feature extraction and for modeling the variations. Although many good systems have been developed, they are not able to perform near what the human can. For instance, as noted by many researchers and as alluded to in the introduction, in addition to pre-processing, segmentation, and feature selection, the handwriting recognition system's performance is heavily dependent on the lexicon size [99]. Usually, the larger the lexicon size the lower the recognition rate which is contrary to our ability to recognize words: the more words we know the better we are at recognizing words. In order for the recognizer to perform at a higher level, we must rely on techniques in computational intelligence; the recognizer should be built using more than traditional statistical and estimation mechanisms. Recently some researchers have used fuzzy logic, neural networks, and rough set in handwritten word recognition [17, 91, 136, 33].

Since the cognitive process of recognizing handwritings is a highly *intelligent* activity, traditional computational methods and models are no longer effective. When dealing with intelligence, we have reached the so-called *conceptual* bottle neck at which in order to make reasonable progress we have to develop new concepts (theories). Given the current set of mathematical tools, or the lack of it, in order to serve our purpose, we have to use a hybrid of unconventional techniques. It has been demonstrated that in many applications where an intelligent decision must be made, soft computing may play a major role. For word recognition for instance, if we consider the parts in the word as the building blocks, we can then use fuzzy sets for modeling their features, fuzzy integral for measuring matching, and machine learning to generate rules for recognition. In this way, we can develop a recognition system that mimics the way we normally try to *figure out* handwritings. And we know how hard it is to recognize the prescription written by a doctor. Based on the idea of building blocks we may also use techniques in genetic algorithms and evolutionary computation to develop handwritten word recognition systems. In order to overcome the limitations of the current recognition systems and to develop the next generation of super recognizers, we need to take a new look at what soft computing can offer.

9.3 References

[1] I.S.I. Abuhaiba and P. Ahmed, "A Fuzzy graph theoretic approach to recognize the totally unconstrained handwritten numerals," *Pattern Recognition*, Vol.26, No.9, pp.1335-1350, 1993.

[2] I.S.I. Abuhaiba, S. Datta and M.J.J. Holt, "Fuzzy state machines to recognize totally unconstructed handwritten strokes," *Image and Vision Computing*, Vol.13, No.10, pp.755-769, December 1995.

[3] K. Arbter, "Affine-invariant Fourier descriptors," in *From Pixels to Features*, J.C. Simon (ed.), Elsevier Science, pp.153-164, 1989.

[4] C. Arcelli, "A Thinning algorithm based on prominence detection," *Pattern Recognition*, Vol.13, No.3, pp.225-235, 1981.

[5] C. Arcelli and G. Sanniti di Baja, "On the sequential approach to medial line transformation," *IEEE Trans. on System Man and Cybernetics*, Vol.8, No.2, pp.139-144, 1978.

[6] N. Arica and F.T. Yarman-Vural, "Optical character recognition for cursive handwriting," *IEEE Trans. on Pattern Analysis and Machine Intelligence*, Vol.24, No.6, pp.801-813, June 2002.

[7] R. Bajcsy, L. Lieberman, "Texture gradient as a depth cue," *Computer Graphics Image Processing*, Vol.5, pp.52-67, 1976.

[8] J. Baker, *Stochastic Modeling as a Means of Automatic Speech Recognition*, Ph.D. thesis, Department of Computer Science, Carnegie-Mellon University, 1975.

[9] J. Baker, "The Dragon system-an overview," *IEEE Trans. on Acoustics, Speech and Signal Processing*, Vol.23, No.1, pp.24-29, February 1975.

[10] L.E. Baum and T. Petrie, "Statistical inference for probabilistic functions of finite state Markov chains," *Annals of Mathematical Statistics*, Vol.37, pp.1554-1563, 1966.

[11] L.E. Baum, T. Petrie, G. Soules and N. Weiss, "A Maximization technique in the statistical analysis of probabilistic functions of Markov chains," *Annals of Mathematical Statistics*, Vol.41, pp.164-171, 1970.

[12] L.E. Baum and J.A. Pagon, "An Inequality with applications to statistical estimation for probabilistic functions of Markov processes and to a model for ecology," *Bulletin of the American Mathematical Society*, Vol.73, pp.360-363, 1967.

[13] J.E. Besag, "Spatial interaction and statistical analysis of lattice systems (with discussion)," *Journal of the Royal Statistical Society (B)*, Vol.36, No.2, pp.192-236, 1974.

[14] J.E. Besag, "On the statistical analysis of dirty pictures (with discussion)," *Journal of the Royal Statistical Society (B)*, Vol.48, No.3, pp.259-302, 1986.

[15] R.K. Belew, J. Mclnerney and N.N. Schraudolph, "Evolving networks: using the genetic algorithm with connectionist learning," *Artificial Intelligence II*, Eds, C.G. Laugton, C. Taylor, J.D. Farmer and S. Rasmussen, Addison-Wesley, pp.511-547, 1991.

[16] S.O. Belkasim, M. Shridhar and M. Ahmadi, "Pattern recognition with moment invariants: a comparative study and new results," *Pattern Recognition*, Vol.24, No.12, pp.1117-1138, 1991.

[17] J.C. Bezdek, J.M. Keller, R. Krishnapuram, and N.R. Pal, *Fuzzy Models and Algorithms on Pattern Recognition and Image Processing*, Kluwer, Norwell MA, 1999.

[18] W.F. Bischof, T. Caelli, "Learning structural descriptions of patterns: a new technique for conditional clustering and rule generation," *Pattern Recognition*, Vol.27, No.5, pp.689-698, 1994.

[19] W.F. Bischof, T. Caelli, "Visual learning of patterns and objects," *IEEE Trans. on Systems, Man, and Cybernetics*, Vol.27, No.6, pp.907-917, 1997.

[20] J.L. Blue, G.T. Candela, P.J. Grother, R. Chellappa and C.L. Wilson, "Evaluation of pattern classifiers for fingerprint and ORC applications," *Pattern Recognition*, Vol.27, No.4, pp.485-501, 1994.

[21] M. Bokser, "Omnidocument technologies," *Proceedings of the IEEE*, Vol.80, No.7, pp.1066-1078, July 1992.

[22] P. Bonissone, "Soft computing: the convergence of emerging reasoning technologies," *Soft Computing*, Vol.1, pp.6-18, 1997.

[23] A.C. Bovik, M. Clark, W.S. Geisler, "Multichannel texture analysis using localized spatial filters," *IEEE Trans. on Pattern Analysis and Machine Intelligence*, Vol.12, No.1, pp.55-73, 1990.

[24] A.C. Bovik, "Analysis of multichannel narrow-band filters for image texture segmentation," *IEEE Trans. on Signal Processing*, Vol.39, No.9, pp.2025-2043, 1991.

[25] A.C. Bovik, N. Gopal, T. Emmoth, A. Restrepo, "Localized measurement of emergent image frequencies by Gabor wavelets," *IEEE Trans. on Information Theory*, Vol.38, No.2, pp.691-712, 1992.

[26] R.M. Bozinovic and S.N. Srihari, "Off-line cursive script word recognition," *IEEE Trans. on Pattern Analysis and Machine Intelligence*, Vol.11, No.1, pp.68-83, January 1989.

[27] M.K. Brown, S. Ganapathy, "Preprocessing techniques for cursive script word recognition," *Pattern Recognition*, Vol.16, No.5, pp.447-458, 1983.

[28] R.M. Brown, T.H. Fay, C.L. Walker, "Handprinted symbol recognition system," *Pattern Recognition*, Vol.21, No.2, pp.91-118, 1988.

[29] S. Brush, "History of the lenz-ising model," *Reviews of Modern Physics*, Vol.39, pp.883-893, 1967

[30] R. Buse, Z.Q. Liu, "Feature extraction and analysis of handwritten words in grey-scale images using Gabor filters," *1st IEEE Int. Conf. on Image Processing*, Vol.1, pp.164-168, 1994.

[31] R. Buse, Z.Q. Liu, T. Caelli, "Using Gabor filters to measure pattern part features and relations," *Pattern Recognition*, Vol.35, No.1, pp.615-625, April 1996.

[32] R. Buse, Z.Q. Liu and T. Caelli, "A Structural and relational approach to handwritten word recognition," *IEEE Trans. on Systems, Man, and Cybernetics*, Part B, Vol.27, No.5, pp.847-861, October 1997.

[33] R. Buse, Z.Q. Liu and J. Bezdek, "Word recognition using fuzzy logic," *IEEE Trans. on Fuzzy Systems*, Vol.10, No.1, pp.65-76, 2002.

[34] J. Cai and Z.Q. Liu, "An Adaptive approach to robust speech recognition," *Proceedings of the Sixth Australian International Conference on Speech Science and Technology*, pp.91-96, Adelaide, December 1996.

[35] J. Cai and Z.Q. Liu, "Markov random field models for handwritten word recognition," *IEEE first International Conference on Intelligent Processing Systems*, Beijing, China, pp.1400-1404, October 28-31, 1997.

[36] J. Cai and Z.Q. Liu, "Integration of structural and statistical information for unconstrained handwritten numeral recognition," *Proceedings of the 14th International Conference on Pattern Recognition*, pp.378-380, Brisben, Australia, August 1998.

[37] J. Cai and Z.Q. Liu, "Integration of structural and statistical information for unconstrained handwritten numeral recognition," *IEEE Trans. on Pattern Analysis and Machine Intelligence*, Vol.21, No.3, pp.263-270, March 1999.

[38] J. Cai and Z.Q. Liu, "Off-line unconstrained handwritten word recognition," *International Journal of Pattern Recognition and Artificial Intelligence*, Vol.14, No.3, pp.259-280, 2000.

[39] J. Cai and Z.Q. Liu, "Hidden Markov models with spectral features for pattern recognition," *IEEE Trans. on Pattern Analysis and Machine Intelligence*, Vol.23, No.12, pp.1454-1458, Dec. 2001.

[40] J. Cai and Z.Q. Liu, "Feature extraction for handwriting recognition using Gabor filters," submitted to *International Journal on Document Analysis and Recognition*, 2001.

[41] J. Cai and Z.Q. Liu, "Pattern recognition using Markov random field models," *Pattern Recognition*, Vol.33, No.3, pp.725-735, March 2002.

[42] J. Canny, "A Computational approach to edge detection," *IEEE Trans. on Pattern Analysis and Machine Intelligence*, Vol.8, No.6, pp.679-698, November 1986.

[43] R.G. Casey and E. Lecolinet, "A Survey of methods and strategies in character segmentation," *IEEE Trans. on Pattern Analysis and Machine Intelligence*, Vol.18, No.7, pp.690-706, July 1996.

[44] D.P. Casasent, J.S. Smokelin, A. Ye, "Wavelet and Gabor Transforms for Detection," *Optical Engineering*, Vol.31, No.9, pp.1893-1898, 1992.

[45] CEDAR CDROM 1, *USPS Office of Advanced Technology*, CEDAR, SUNY at Buffalo, 1992.

[46] V. Cerny, "Thermodynamical approach to the traveling salesman problem: an efficient simulated algorithm," *Journal of Optimization Theory and Applications*, Vol.45, pp.41-51, 1985.

[47] R. Chellappa and S. Chatterjee, "Classification of textures using Gaussian Markov random fields," *IEEE Trans. on Acoustics, Speech and Signal Processing*, Vol.33, pp.959-963, 1985.

[48] M.Y. Chen, A. Kundu, and S.N. Srihari, "Variable duration hidden Markov model and morphological segmentation for handwritten word recognition," *IEEE Trans. on Image Processing*, Vol.4, No.12, pp.1675-1688, December 1995.

[49] M.Y. Chen, A. Kundu, and J. Zhou, "Off-line handwritten word recognition using a hidden Markov model type stochastic network," *IEEE Trans. on Pattern Analysis and Machine Intelligence*, Vol.16, pp.481-496, 1994.

[50] S.J. Chen, C.L. Hwang, and F.P. Hwang, *Fuzzy Multiple Attribute Decision Making, Methods and Applications*, Springer-Verlag, Berlin, 1992.

[51] D. Cheng and H. Yan, "Recognition of handwritten digits based on contour information." *Pattern Recognition*, Vol.31, No.3, pp.135-255, 1998.

[52] J.H. Chiang and P.D. Gader, "Hybrid fuzzy-neural systems in handwritten word recognition," *IEEE Trans. on Fuzzy Systems*, Vol.5, No.4, pp.497-510, November 1997.

[53] R. Chellappa and R. Bagdazian, "Fourier coding of image boundaries," *IEEE Trans. on Pattern Analysis and Machine Intelligence*, Vol.6, No.1, pp.102-105, January 1984.

[54] F.H. Cheng and W.H. Hsu, "Research on Chinese ORC in Taiwan," in *Character and Handwriting Recognition: Expanding Frontiers*, Ed, P.S.P. Wang, World Scientific Series in Computer Science, Vol.30, pp.139-164, 1991.

[55] F.H. Cheng and W.H. Hsu, "Parallel algorithm for corner finding on digital curves," *Pattern Recognition Letters*, Vol.8, pp.47-53, 1988.

[56] F.H. Cheng, "Multi-stroke relaxation matching method for handwritten Chinese character recognition," *Pattern Recognition*, Vol.31, No.4, pp.401-410, 1998.

[57] M. Clark, A.C. Bovik, and W.S. Geisler, "Texture segmentation using a class of narrowband filters," *Proc. Int. Conf. on Acoustics Speech and Signal Processing*, pp.571-574, 1987.

[58] S.B. Cho and J.H. Kim, "Multiple network fusion using fuzzy logic," *IEEE Trans. on Neural Networks*, Vol.6, No.2, pp.497-501, March 1995.

[59] S.B. Cho, "Neural-network classifiers for recognizing totally unconstrained handwritten numerals," *IEEE Trans. on Neural Networks*, Vol.8, No.1, pp.43-53, January 1997.

[60] W. Cho, S.W. Lee, and J.H. Kim, "Modeling and recognition of cursive words with hidden Markov models," *Pattern Recognition*, Vol.12, No.12, pp.1941-1953, 1995.

[61] P.B. Chou and C.M. Brown, "The Theory and practice of Bayesian image labeling," *International Journal of Computer Vision*, Vol.4, pp.185-210, 1990.

[62] P.B. Chou, P.R. Cooper, M.J. Swain, C.M. Brown, and L.E. Wixson, "Probabilistic network inference for cooperative high and low level vision," Chapter 9, in *Markov Random Fields: Theory and Applications*, R. Chellappa and A. Jain, Eds., Academic Press, Boston, pp.211-244, 1993.

[63] P. Chou, T. Lookabaugh and R. Gray, "Entropy-constrained vector quantization," *IEEE Trans. on Acoustics, Speech and Signal Processing*, Vol.37, pp.31-42, 1989.

[64] S.L. Chou and W.H. Tsai, "Recognizing handwritten Chinese characters by stroke-segment matching using an iteration scheme," in *Character and Handwriting Recognition: Expanding Frontiers*, Ed, P.S.P. Wang, World Scientific Series in Computer Science, Vol.30, pp.175-197, 1991.

[65] M. Das, M.J. Paulik and N.K. Loh, "A Bivariate autoregressive modeling technique for analysis and classification of planar shapes," *IEEE Trans. on Pattern Analysis and Machine Intelligence*, Vol.12, No.1, pp.97-103, January 1990.

[66] J.G. Daugman, "Six formal properties of two-dimensional anisotropic visual filters: structural principles and frequency/orientation selectivity," *IEEE Trans. on Systems Man and Cybernetics*, Vol.13, No.5, pp.882-887, 1983.

[67] J. G. Daugman, "Uncertainty relation for resolution in space, spatial frequency, and orientation optimized by two-dimensional visual cortical filters," *Journal of the Optical Society of America*, Vol.2, No.7, pp.1160-1169, July 1985.

[68] J.G. Daugman, "Complete discrete 2-D Gabor transformations by neural networks for image analysis and compression," *IEEE Trans. on Acoustics Speech and Signal Processing*, Vol.36, No.7, pp.1169-1179, 1988.

[69] J.R. Deller, J.G. Proakis and J.H.L. Hansen, *Discrete-time Processing of Speech Signals*, New York: Maxwell Macmillan, 1993.

[70] H. Derin and P.A. Kelly, "Discrete-index Markov-type random processes," *Proceedings of the IEEE*, Vol.77, No.10, pp.1485-1510, October 1989.

[71] P.A. Devijver and J. Kittler, *Pattern Recognition : a statistical approach*, Prentice Hall: Englewood Cliffs, 1982.

[72] A.P. Dempster, N.M. Laird and D.B. Rubin, "Maximum likelihood from incomplete data via EM algorithm," *Journal of the Royal Statistical Society, Series B (methodological)* , Vol.39, No.1, pp.1-38, 1977.

[73] H. Drucker, "Fast decision tree ensembles for optical character recognition," *Proceedings of the Fifth Annual Symposium on Document Analysis and Information Retrieval*, pp.137-147, 1996.

[74] S.R. Dubois and F.H. Glanz, "An Autoregressive model approach to two-dimensional shape classification," *IEEE Trans. on Pattern Analysis and Machine Intelligence*, Vol.8, No.1, pp.55-66, January 1986.

[75] J.M. Du Buf, "Gabor phase in texture discrimination," *Signal Processing*, Vol.21, pp.221-240, 1990.

[76] J.M. Du Buf, P. Heitkamper, "Texture features based on Gabor phase," *Signal Processing*, Vol.23, pp.227-244, 1991.

[77] M. Eden, M. Halle, "The Characterization of cursive writing," *Proc. 4th London Symposium on Information Theory*, pp.287-299, 1961.

[78] M. Eden, "Handwriting generalization and recognition," in *Recognizing Patterns*, Eds, Kolers and M. Eden, M.I.T. Press, Cambridge, pp.138-154, 1968.

[79] R.J. Elliott, L. Aggoun and J.B. Moore, *Hidden Markov Models: estimation and control*, Springer-Verlag, New York, 1995.

[80] A.J. Elms, *The Representation and Recognition of Text Using Hidden Markov Models*, Ph.D. thesis, Department of Electronic and Electrical Engineering, University of Surrey, UK, 1996.

[81] A.J. Elms and J. Illingworth, "Combination of HMMs for the representation of printed characters in noisy document images," *Image and Vision Computing*, Vol.13, No.5, pp.385-392, June 1995.

[82] R. Farag, "Word-level recognition of cursive script," *IEEE Trans. on Comput.*, Vol.28, pp.172-175, Feb. 1979.

[83] I. Fogel, D. Sagi, "Gabor filters as texture discriminators," *Biological Cybernetics*, Vol.61, pp.103-113, 1989.

[84] Y. Freund and R.E. Schapire, "Experiments with a new boosting algorithm," *Proceedings of the 13th International Conference on Machine Learning*, pp.148-156, 1996.

[85] L.S. Frishkopf, L.D. Harmon, "Machine reading of cursive script," *Proc. 4th London Symposium on Information Theory*, pp.300-316, 1961.

[86] H. Freeman, "Boundary encoding and processing," in: *Picture Processing and Psychopictorics*, B.S. Lipkin and A. Rosenfeld, Eds., New York: Academic Press, pp.241-306, 1970.

[87] K.S. Fu and T.L. Booth, "Grammatical inference: introduction and survey-Part I," *IEEE Trans. on Pattern Analysis and Machine Intelligence*, Vol.8, No.3, pp.343-359, May 1986.

[88] K.S. Fu and T.L. Booth, "Grammatical inference: introduction and survey-Part II," *IEEE Trans. on Pattern Analysis and Machine Intelligence*, Vol.8, No.3, pp.360-375, May 1986.

[89] K.S. Fu, *Syntactic Pattern Recognition and Applications*, Prentice-Hall, Englewood Cliffs, N.J., 1982

[90] D. Gabor, "Theory of communications," *Journal of The Institute of Electrical Engineering*, Vol.93, pp.429-457, 1946.

[91] P. Gader, J. Keller, and J. Cai, "A Fuzzy logic system for the detection and recognition of street number fields on handwritten postal addresses," *IEEE Trans. on Fuzzy Systems*, Vol.3, No.1, pp.83-95, 1995.

[92] D. Gader, M. Mohammed, and J.H. Chiang, "Handwritten word recognition with character and inter character neural networks," *IEEE Trans. System, Man, and Cybernetics*, Vol.27, No.1, pp.158-164, 1997.

[93] M.D. Garris, R.A. Wilkinson, C.L. Wilson, "Analysis of a biologically motivated neural network for character recognition," *Proc. Analysis of Neural Network Applications*, pp.160-175, 1991.

[94] S. Geman and D. Geman, "Stochastic relaxation, Gibbs distribution, and the Bayesian restoration of images," *IEEE Trans. on Pattern Analysis and Machine Intelligence*, Vol.6, No.6, pp.721-741, November 1984.

[95] D. Geman, S. Geman, C. Graffigne and P. Dong, "Boundary detection by constrained optimization," *IEEE Trans. on Pattern Analysis and Machine Intelligence*, Vol.12, No.7, pp.609-628, July 1990.

[96] D. Geman and G. Reynolds, "Constrained restoration and the recovery of discontinuities," *IEEE Trans. on Pattern Analysis and Machine Intelligence*, Vol.14, No.3, pp.367-383, March 1992.

[97] D. Geman, G. Reynolds and C. Yang, "Stochastic algorithms for restricted image spaces and experiments in deblurring," Chapter 3, in *Markov Random Fields: Theory and Application*, R. Chellappa and A. Jain, Eds., Academic Press, Boston, pp.39-68, 1993.

[98] R.C. Gonzalez, P. Wintz, *Digital Image Processing*, 2nd ed., Addison-Wesley, Reading, Massachusetts, 1987.

[99] V. Govindraraju, P. Slavik, and H.H. Xue, "Use of lexicon density in evaluating word recognizers," *IEEE Trans. on Pattern Analysis and Machine Intelligence*, Vol.24, No.6, pp.789-800, June 2002.

[100] R.M. Gray, "Vector quantization," *IEEE Acoustics, Speech, and Signal Processing Magazine*, Vol.1, pp.4-29, April 1984.

[101] W.E. Griffiths, H. Lutkepohl and M.E. Bock, *Readings in Econometric Theory and Practice: a volume in honor of George Judge*, Amsterdam; New York : North-Holland, 1992.

[102] D. Guillevic, C.Y. Suen, "Cursive script recognition: a fast reader scheme," *Proc. Int. Conf. on Document Analysis and Recognition*, pp.311-314, 1993.

[103] T.M. Ha and H. Bunke, "Off-line, handwritten numeral recognition by perturbation method," *IEEE Trans. on Pattern Analysis and Machine Intelligence*, Vol.19, No.5, pp.535-539, May 1997.

[104] K.C. Hayes, "Reading handwritten words using hierarchical relaxation," *Computer Graphics and Image Processing*, Vol.14, No.3, pp.344-364, September 1980.

[105] R.M. Haralick, "Digital step edges from zero-crossings of second directional derivatives," *IEEE Trans. on Pattern Analysis and Machine Intelligence*, Vol.6, No.1, pp.58-68, January 1984.

[106] E.R. Hancock, and J. Kittler, "Edge labelling using dictionary based relaxation," *IEEE Trans. on Pattern Analysis and Machine Intelligence*, Vol.12, No.2, pp.161-182, 1990.

[107] S. Hashem and B. Schmeiser, "Improving model accuracy using optimal linear combinations of trained neural networks," *IEEE Trans. on Neural Networks*, Vol.6, No.3, pp.792-794, May 1995.

[108] D.J. Heeger, "Model for the extraction of image flow," *Opt. Soc. Am. A*, Vol.4, No.8, pp.1455-1471, 1987.

[109] T.K. Ho, J.J. Hull, S.N. Srihari, "A Word shape analysis approach to lexicon based word recognition," *Pattern Recognition Letters*, Vol.13, pp.821-826, 1992.

[110] T.K. Ho, J.J. Hull and S.N. Srihari, "Decision combination in multiple classifier systems," *IEEE Trans. on Pattern Analysis and Machine Intelligence*, Vol.16, No.1, pp.66-75, January 1994.

[111] X.D. Huang, Y. Ariki and M.A. Jack, *Hidden Markov Models for Speech Recognition*, Edinburgh University Press, 1990.

[112] X.D. Huang, M.A. Jack and Y. Ariki, "Parameter re-estimation of semi-continuous hidden Markov models with feedback to vector quantisation codebook," *IEE Electronics Letters*, Vol.24, No.24, pp.1375-1377, 1988.

[113] X.D. Huang and M.A. Jack, "Semi-continuous hidden Markov models for speech recognition," *Computer Speech and Language*, Vol.3, pp.239-251, 1989.

[114] J. Hu, M.K. Brown and W. Turin, "HMM based on-line handwriting recognition," *IEEE Trans. on Pattern Analysis and Machine Intelligence*, Vol.18, No.10, pp.1039-1045, October 1996.

[115] M.K. Hu, "Visual pattern recognition by moment invariants," *IRE Trans. on Information Theory*, Vol.8, pp.179-187, February 1962.

[116] J.J. Hull, "A Database for handwritten text recognition research," *IEEE Trans. on Pattern Analysis and Machine Intelligence*, Vol.16, No.5, pp.550-554, May 1994.

[117] S.H.Y. Hung and T. Kasvand, "Critical points on a perfectly 8-6- connected thin binary line," *Pattern Recognition*, Vol.16, No.3, pp.297-306, 1983.

[118] Q. Huo and C. Chan, "A Study on the use of bi-directional contextual dependence in Markov random field-based acoustic modelling for speech recognition," *Computer Speech and Language*, Vol.10, pp.95-105, 1996.

[119] C.L. Hwang, K. Yoon, *Multiple Attribute Decision Making, Methods and Applications: A State-of-the-Art-Survey*, Springer-Verlag, Berlin, 1981.

[120] Y.S. Hwang, and S.Y. Bang, "Recognition of unconstrained handwritten numerals by a radial basis function neural network classifier," *Pattern Recognition Letters*, Vol.18, pp.657-664, 1997.

[121] R. Hummel and S. Zucker, "On the foundations of relaxation labeling processing," *IEEE Trans. On Pattern Analysis and Machine Intelligence*, Vol.5, No.3, pp.267-287, May 1983.

[122] S. Impedovo, L. Ottaviano and S. Occhinegro, "Optical character recognition-a survey," in *Character and Handwriting Recognition*, Ed, P.S.P. Wang, Word Scientific Series in Computer Science, Vol.30, pp.1-24, 1991.

[123] A.M. Jacobs, J. Grainger, "Models of visual word recognition - sampling the state of the art," *Journal of Experimental Psychology*, Vol.20, No.6, pp.1311-1334, 1994.

[124] A.K. Jain, F. Farrakhina, "Unsupervised texture segmentation using Gabor filters," *Pattern Recognition*, Vol.24, No.2, pp.1167-1186, 1991.

[125] A.K. Jain, S.K. Bhattacharjee, "Text segmentation using Gabor filters for automatic document processing," *Machine Vision and Applications*, Vol.5, No.3, pp.169-184, 1992.

[126] A.K. Jain, S.K. Bhattacharjee, "Address block location on envelopes using Gabor filters," *Pattern Recognition*, Vol.25, No.12, pp.1459-1477, 1992.

[127] F. Jelinek, "Continuous speech recognition by statistical methods," *Proceedings of the IEEE*, Vol.64, pp.532-556, April 1976.

[128] F.C. Jeng and J.W. Wood, "Compound Gauss-Markov random fields for image estimation," *IEEE Transactions on Signal Processing*, Vol.39, No.3, pp.683-679, March 1991.

[129] X.Y. Jinag and H. Bunke, "Simple and fast computation of moments," *Pattern Recognition*, Vol.24, No.8, pp.801-806, 1991.

[130] J.P. Jones and L.A. Palmer, "An Evaluation of the two-dimensional Gabor filter model of simple receptive fields in the cat striate cortex," *Journal of Neurophysiology*, Vol.58, pp.1233-1258, 1987.

[131] B.H. Juang, "Maximum likelihood estimation for mixture multivariate stochastic observations of Markov chains," *AT&T System Technical Journal*, Vol.64, pp.1235-1249, July-August 1985.

[132] B.H. Juang, S.E. Levinson and M.M. Sondhi, "Maximum likelihood estimation for multivariate mixture observations of Markov chains," *IEEE Trans. on Information Theory*, Vol.32, pp.307-309, March 1986.

[133] N. N. Karnik, J. M. Mendel and Q. Liang "Type-2 Fuzzy Logic Systems," *IEEE Trans. on Fuzzy Systems*, vol. 7, pp. 643-658, Dec. 1999.

[134] Y. Kato and M. Yasuhara, "Recovery of drawing order from single-stroke handwriting images," *IEEE Trans. on Pattern Analysis and Machine Intelligence*, Vol.22, No.9, pp.938-949, Sept. 2000.

[135] H. Kauppinen, T. Seppänen and M. Pietikäinen, "An Experimental comparison of autoregressive and Fourier-based descriptors in 2-D shape classification," *IEEE Trans. on Pattern Analysis and Machine Intelligence*, Vol.17, pp.201-207, 1995.

[136] D.J. Kim and S.Y. Bang, "A Handwritten numeral character classification using tolerant rough set," *IEEE Trans. on Pattern Analysis and Machine Intelligence*, Vol.22, No.9, pp.923-937, Sept. 2000.

[137] H.J. Kim, K.H. Kim, S.K. Kim and J.K. Lee, "On-line recognition of handwritten Chinese characters based on hidden Markov models," *Pattern Recognition*, Vol.30, No.9, pp.1489-1500, 1997.

[138] F. Kimura, M. Shridhar, Z. Chen, "Improvements of a lexicon directed algorithm for recognition of unconstrained handwritten words," *Proc. Int. Conf. on Document Analysis and Recognition*, pp.18-22, 1993.

[139] G. Kim and V. Govindaraju, "A Lexicon driven approach to handwritten word recognition for real-time applications," *IEEE Trans. on Pattern Analysis and Machine Intelligence*, Vol.19, No.4, pp.366-379, April 1997.

[140] R. Kindermann and J.L. Snell, *Markov random fields and their applications*, Rhode Island: American Mathematical Society, 1980.

[141] S. Kirkpatrick, C.D. Gellate and M.P. Vecchi, "Optimization by simulated annealing," *Science*, Vol.220, pp.671-680, 1983.

[142] J. Kittler and J. Föglein, "Contextual decision rules for objects in lattice configurations," *Proceedings of the 7th International Conference on Pattern Recognition*, Montreal, pp.270-272, 1984.

[143] J. Kittle, M. Hatef, R.P.W. Duin and J. Matas, "On combining classifiers," *IEEE Trans. on Pattern Analysis and Machine Intelligence*, Vol.20, No.3, pp.226-239, March 1998.

[144] J. Kittler, "Relaxation labeling," in *Pattern Recognition Theory and Applications*, P.A. Devijver and J. Kittler, Ed., NATO ASI Series,Vol.F30, pp.99-108, 1987.

[145] J. Kittle, "Combining classifiers: a theoretical framework," *Pattern Analysis and Applications*, Vol.1, No.1, pp.18-27, 1998.

[146] U. Kjærulff, "Optimal decomposition of probabilistic networks by simulated annealing," *Statistics and Computing*, Vol.2, pp.7-17, 1991.

[147] T. Kohonen, "The Self-organizing map," *Proceedings of the IEEE*, Vol.78, No.9, pp.1464-1480, 1990.

[148] F.P. Kuhl and C.R. Giardina, "Elliptic Fourier features of a closed contour," *Computer Vision, Graphics and Image processing*, Vol.18, pp.236-258, 1982.

[149] S.S. Kuo and O.E. Agazzi, "Keyword spotting in poorly printed documents using pseudo 2-D hidden Markov models," *IEEE Trans. on Pattern Analysis and Machine Intelligence*, Vol.16, No.8, pp.842-848, August 1994.

[150] L. Lam, S.W. Lee, and C.Y. Suen, "Thinning methodologies-a comprehensive survey," *IEEE Trans. on Pattern Analysis and Machine Intelligence*, Vol.14, No.9, pp.869-885, September 1992.

[151] Y. LeCun, I. Guyon, L.D. Jackel, D. Henderson, B. Boser, R.E. Howard, J.S. Denker, W. Hubbard, and H.P. Graf, "Handwritten digit recognition: applications of neural network chips and automatic learning," *IEEE Communications Magazine*, Vol.11, pp.41-64, 1989.

[152] E. Lecolinet, J.V. Moreau, "Off-line recognition of handwritten cursive script for automatic reading of city names on real mail," *Proc. 10th Int. Conf. on Pattern Recognition*, vol.1, pp.674-675, 1990.

[153] K.F. Lee, *Large-vocabulary Speaker Independent Continuous Speech Recognition: The SPINX system*, PH.D. Thesis, Computer Science Department, Carnegie Mellon University, 1988.

[154] S.W. Lee, "Off-line recognition of totally unconstrained handwritten numerals using multilayer cluster neural network," *IEEE Trans. on Pattern Analysis and Machine Intelligence*, Vol.188 No.6, pp.648-652, June 1996.

[155] H.C. Lee and K.S. Fu, "A Stochastic syntax analysis procedure and its application to pattern classification," *IEEE Trans. on Computers*, Vol.21, pp.660-666, July 1972.

[156] S.W. Lee and Y.J. Kim, "Direct extraction of topographic features for gray scale character recognition," *IEEE Trans. on Pattern Analysis and Machine Intelligence*, Vol.17, No.7, pp.724-729, July 1995.

[157] S.E. Levinson, "Continuously variable duration hidden Markov models for automatic speech recognition," *Computer, Speech and Language*, Vol.1, No.1, pp.29-45, March 1986.

[158] E. Levin and R. Pieraccini, "Dynamic planar warping for optical character recognition," *Proceedings of International Conference on Acoustics, Speech and Signal Processing*, San Francisco, USA, pp.23-26, 1992.

[159] S.Z. Li, *Markov Random Field Modeling in Computer Vision*, Tokyo: Springer-Verlag, 1995.

[160] S.Z. Li, "Parameter estimation for optimal object recognition: theory and application," *International Journal of Computer Vision*, Vol.21, pp.207-222, 1997.

[161] L.A. Liporace, "Maximum likelihood estimation for multivariate observations of Markov sources," *IEEE Trans. on Information Theory*, Vol.2, pp.729-734, September 1982.

[162] C.L. Liu, M. Koga, and H, Fujisawa, "Lexicon-driven segmentation and recognition of handwritten character strings for Japanese address reading," *IEEE Trans. on Pattern Analysis and Machine Intelligence*, Vol.24, No.11, pp.1425-1437, Nov. 2002.

[163] Z.Q. Liu, "Bayesian paradigms in image processing," *International Journal of Pattern Recognition and Artificial Intelligence*, Vol.11, No.1, pp.3-33, February 1997.

[164] Z.Q. Liu, R.M. Rangayyan, C.B. Frank, "A New image analysis method for collagen alignment in knee ligaments using Gabor filters," *Proc. 3rd Int. Symposium on Computer Based Medical Systems*, pp.68-74, 1990.

[165] B.T. Lowerre and R. Reddy, "HARPY speech understanding system," *Trends in speech recognition*, Eds, W.A. Lea, pp.340-360, Prentice-Hall, Englewood Cliffs, N.J., 1980.

[166] Y. Lu and M. Shridhar, "Character segmentation in handwritten words: an overview," *Pattern Recognition*, Vol.29, No.1, pp.77-96, January 1996.

[167] S. Madhvanath and V. Govindaraju, "The Role of holistic paradigms in handwritten word recognition," *IEEE Trans. on Pattern Analysis and Machine Intelligence*, Vol.23, No.2, pp.149-164, Feb. 2001.

[168] J. Makhoul, T. Starner, R. Schartz and G. Lou, "On-line cursive handwriting recognition using speech recognition models," *Proceedings of the International Conference on Acoustics, Speech and Signal Processing*, pp.125-128, Adelaide, Australia, April 1994.

[169] J. Makhoul, S. Roucos and H. Gish, "Vector quantization in speech coding," *Proceedings of the IEEE*, Vol.73, pp.1551-1588, November 1985.

[170] A. Malaviya and L. Peters, "Fuzzy features description of handwriting patterns," *Pattern Recognition*, Vol.30, No.10, pp.1592-1604, 1997.

[171] S. Marcelja, "Mathematical description of the responses of simple cortical cells," *Journal of the Optical Society of America*, Vol.70, No.11, pp.1297-1300, 1980.

[172] D.J. Markel and A.H. Gray, *Linear Prediction of Speech*, New York: Springer-Verlag, 1976.

[173] D.C. Marr and E.C. Hildreth, "Theory of edge detection," *Proceedings of the Royal Society of London*, Series B, Vol.207, pp.187-217, 1980.

[174] L. Mason, P. Bartlett, and J. Baxter, "Direct optimization of margins improves generalization in combined classifiers," *Technical Report*, Department of Systems Engineering, Australian National University, 1998.

[175] R. Mehrotra, K.R. Namuduri, and N. Ranganathan, "Gabor filter-based edge detection," *Pattern Recognition*, Vol.25, No.12, pp.1479-1494, 1992.

[176] J.M. Mendel, "Fuzzy logic systems for engineering: a tutorial," *Proceedings of IEEE*, Vol.83, pp.345-377, 1995.

[177] N. Metropolis, A. Rosenbluth, M. Rosenbluth, A. Teller, and E. Teller, "Equations of state calculations by fast computational machine," *Journal of Chemical Physics*, Vol.21, pp.1087-1091, 1953.

[178] O.R. Mitchell and T.A. Grogan, "Global and partial shape discrimination for computer vision," *Optical Engineering*, Vol.23, pp.484-491, 1984.

[179] J.W. Modestino and J.Zhang, "A Markov random field model-based approach to image interpretation," Chapter 14, in *Markov Random Fields: Theory and Applications*, R. Chellappa and A. Jain, Eds., Academic Press, Boston, pp.369-408, 1993.

[180] S. Mori, C.Y. Suen, and K. Yamamoto, "Historic review of ORC research and development," *Proceedings of IEEE*, Vol.80, No.7, pp.1029-1058, 1992.

[181] C.S. Myers and L.R. Rabiner, "A Level building dynamic time warping algorithm for connected word recognition," *IEEE Trans. on Acoustics, Speech, and Signal Processing*, Vol.29, No.2, pp.284-297, 1981.

[182] K.S. Nathan, H.S.M. Beigi, J. Subrahmonia, G.J. Clary, and H. Maruyama, "Real-time on-line unconstrained handwriting recognition using statistical methods," *Proceedings of the International Conference on Acoustics, Speech and Signal Processing*, pp.2619-2622, Detroit, USA, June 1995.

[183] H. Nishida and S. Mori, "An Algebraic approach to automatic construction of structural models," *IEEE Trans. on Pattern Analysis and Machine Intelligence*, Vol.15, No.12, pp.1298-1311, December 1993.

[184] H. Nishida, "An Approach to integration of off-line and on-line recognition of handwriting," *Pattern Recognition Letters*, Vol.16, No.11, pp.1213-1219, No.v 1995.

[185] J.R. Norris, "Markov chains,", Chapter one, in *Discrete-time Markov Chains*, Cambridge University Press, 1997.

[186] M. Pötzsch, N. Krügert, and V. Malsburg, "Improving object recognition by transforming Gabor filter responses," *Network: Computation in Neural Systems*, Vol.7, No.2, pp.341-347, May 1996.

[187] N. Otsu, "A Threshold selection method from grey-level histograms," *IEEE Trans. on Systems Man and Cybernetics*, Vol.9, pp.63-66, (1979).

[188] D.K. Panjwani and G. Healey, "Markov random field models for unsupervised segmentation of textured color images," *IEEE Trans. on Pattern Analysis and Machine Intelligence*, Vol.17, No.10, pp.939-954, October 1995.

[189] H.S. Park and S.W. Lee, "A truly 2-D hidden Markov model for off-line handwritten character recognition," *Pattern Recognition*, Vol.31, No.12, pp.1849-1864, 1998.

[190] H.S. Park and S.W. Lee, "Off-line recognition of large-set handwritten characters with multiple hidden Markov models," *Pattern Recognition*, Vol.29, No.2, pp.231-244, 1996.

[191] A. Papoulis, *Probability, Random Variables and Stochastic Processes*, New York: McGraw-Hill, 1984.

[192] T. Paquet and Y. Lecourtier, "Recognition of handwritten sentences using a restricted lexicon," *Pattern Recognition*, Vol.26, No.3, pp.391-407, March 1993.

[193] M. Parizeau, R. Plamondon and G. Lorette, "Fuzzy-shape grammars for cursive script recognition," *Proceedings of International Workshop on Advances in Structural and Syntactic Pattern Recognition*, Ed, H. Bunke, pp.320-332, 1992.

[194] T. Pavlidis, "Why progress in machine vision is so slow," *Pattern Recognition Letters*, Vol.13, pp.221-225, 1992.

[195] T. Pavlidis, "Structural descriptions and graph grammars," in *Pictorial Information Systems*, Eds, S.K. Chang and K.S. Fu, Lecture Notes in Computer Science 80, Springer-Verlag, Berlin Heidelberg, pp.86-103, 1980.

[196] T. Pavlidis, *Structural Pattern Recognition*, Springer-Verlag, Berlin, 1977.

[197] T. Pavlidis and F. Ali, "Computer recognition of handwritten numerals by polygonal approximations," *IEEE Trans. on Pattern Analysis and Machine Intelligence*, Vol.5, No.6, pp.610-614, November 1975.

[198] E. Persoon and K.S. Fu, "Shape discrimination using Fourier descriptors," *IEEE Trans. on Pattern Analysis and Machine Intelligence*, Vol.8, No.3, pp.388-397, May 1986.

[199] J. Picone, "Continuous speech recognition using hidden Markov models," *IEEE Acoustics, Speech, and Signal Processing Magazine*, pp.26-41, July 1990.

[200] R. Plamomdon, W. Guerfali and X. Li, "The Generation of oriental characters: new perspectives for automatic handwriting processing," *International Journal of Pattern Recognition and Artificial Intelligence*, Vol.12, No.1, pp.31-44, February 1998.

[201] R. Plamomdon and S.N. Srihari, "On-line and off-line handwriting recognition: a comprehensive survey," *IEEE Trans. on Pattern Analysis and Machine Intelligence*, Vol.22, No.1, pp.63-84, Jan. 2000.

[202] J.C. Pettier, J. Camillerapp, "Script representation by a generalized skeleton," *Proc. 2nd Int. Conf. on Document Analysis and Representation*, pp.850-853, 1993.

[203] T.E. Portegys, "A Search technique for pattern recognition using relative distances," *IEEE Trans. on Pattern Analysis and Machine Intelligence*, Vol.17, No.9, pp.910-914, 1995.

[204] M. Porat, Y.Y. Zeevi, "Localized Texture Processing in Vision: Analysis and Synthesis in the Gaborian Space," *IEEE Trans. on Biomedical Engineering*, Vol.36, No.1, pp.115-129, 1989.

[205] W.K. Pratt, *Digital Image Processing*, John Wiley & Sons, 1991.

[206] R.J. Prokop and A.P. Reeves, "A Survey of moment-based techniques for unoccluded object representation and recognition," *Computer Vision Graphics and Image Processing (CVGIP): Graphical Models and Image Processing*, Vol.54, No.5, pp.438-460, September 1992.

[207] L.R. Rabiner and B.H. Juang, "An Introduction to hidden Markov models," *IEEE Acoustics, Speech, and Signal Processing Magazine Magazine*, pp.4-16, January 1986.

[208] L.R. Rabiner, B.H. Juang, S.E. Levinson and M.M. Sondhi, "Recognition of isolated digits using hidden Markov models with continuous mixture densities," AT&T Tech. J., Vol.64, No.6, pp.1211-1222, July-Aug. 1986.

[209] L.R. Rabiner, "A Tutorial on hidden Markov model and selected applications in speech recognition," *Proceedings of the IEEE*, Vol.77, pp.257-286, 1989.

[210] L.R. Rabiner and S.E. Levinson, "A Speaker-independent, syntax-directed, connected word recognition system based on hidden Markov models and level building," *IEEE Trans. on Acoustics, Speech, and Signal Processing*, Vol.33, No.3, pp.561-573, 1985.

[211] A.F.R. Rahman and M.C. Fairhurst, "An Evaluation of multi-expert configurations for the recognition of handwritten numerals," *Pattern Recognition*, Vol.31, No.9, pp.1255-1273, 1998.

[212] R.A. Redner and H.F. Walker, "Mixture densities, maximum likelihood and the EM algorithm," *SIAM Review*, Vol.26, pp.195-239, 1984.

[213] T.H. Reiss, "The Revised fundamental theorem of moment invariants," *IEEE Trans. on Pattern Analysis and Machine Intelligence*, Vol.13, No.8, pp.830-834, August 1991.

[214] J.G. Roberts, "Machine perception of three dimensional solids," in *Optical and electro-optical information processing*, J.T. Tippett, Ed., Cambridge, MA: MIT Press, pp.150-197, 1965.

[215] J. Rocha and T. Pavlids, "A Shape analysis model with applications to a character recognition system," *IEEE Trans. on Pattern Analysis and Machine Intelligence*, Vol.16, No.4 pp.393-404, April 1994.

[216] A. Rosenfeld and A.C. Kak, *Digital Picture Processing*, Vol.2, New York: Academic Press, 1982.

[217] A. Rosenfeld, R. Hummel, and S. Zucker, "Scene labeling by relaxation operations," *IEEE Trans. on Systems, Man and Cybernetics*, Vol.6, pp.420-433, 1976.

[218] P.K. Sahoo, S. Soltani. A.K.C. Wong, and Y.C. Chen, "Survey of thresholding techniques," *Computer Vision, Graphics, and Image Processing*, Vol.41, No.2, pp.233-260, 1988.

[219] T.D. Sanger, "Stereo disparity computation using Gabor filters," *Biological Cybernetics*, Vol.59, pp.405-418, 1988.

[220] G. Sanniti di Baja, "Representing shape by line patterns," in *Advances in Structural and Syntactical Pattern Recognition*, Eds, P. Perner, P. Wang and A. Rosenfeld, Lecture Notes in Computer Science 1121, Springer-Verlag, Berlin, pp.231-239, 1996.

[221] R. Schwartz and S. Austin, "A Comparison of several approximate algorithms for finding multiple (N-best) sentence hypotheses," *Proceedings of International Conference on Acoustics, Speech and Signal Processing*, pp.701-704, Toronto, Canada, 1991.

[222] H. Schwenk and Y. Bengio, "Adaptive boosting of neural networks for character recognition," *Technical Report #1072*, Department d'Informatique et Recherche Operationnelle, Universite de Montreal, May 1997.

[223] S. Sclaroff, "Deformable prototypes for encoding shape categories in image databases," *Pattern Recognition*, Vol.30, No.4, pp.627-641, April 1997.

[224] S. Sclaroff and A. Pentland, "Modal matching for correspondence and recognition," *IEEE Trans. on Pattern Analysis and Machine Intelligence*, Vol.17, No.6, pp.545-561, 1995.

[225] I. Sekita, T. Kurita and N. Otsu, "Complex autoregressive model for shape recognition," *IEEE Trans. on Pattern Analysis and Machine Intelligence*, Vol.14, No.4, pp.489-496, April 1992.

[226] B.M. Shahshahani, "A Markov random field approach to Bayesian speaker adaptation," *IEEE Trans. on Speech and Audio Processing*, Vol.5, No.2, pp.183-191, March 1997.

[227] A.W. Senior, F. Fallside, "Off-line Handwriting Recognition by Recurrent Error Propagation Networks," *British Machine Vision Conference*, 1992.

[228] J.C. Simon, "Off-line cursive word recognition," *Proceedings of the IEEE*, Vol.80, No.7, pp.1150-1161, July 1992.

[229] B.K. Sin and J.H. Kim, "Ligature modeling for online cursive script recognition," *IEEE Trans. on Pattern Analysis and Machine Intelligence*, Vol.19, No.6, pp.623-633, June 1997.

[230] H.F. Silverman and D.P. Morgan, "The Application of dynamic programming to connected speech recognition," *IEEE Acoustics, Speech, and Signal Processing Magazine*, Vol.7, pp.6-25, July 1990.

[231] D.F. Specht, "Probabilistic neural networks and general regression neural networks," in: *Fuzzy Logic and Neural Network Handbook*, Ed, C.H. Chen, McGraw-Hill Inc, New York, pp.3.1-3.44, 1996.

[232] A. Speis and G. Healey, "Feature extraction for texture discrimination via random field models with random spatial interaction," *IEEE Trans. on Image Processing*, Vol.5, No.4, pp.635-645, April 1996.

[233] F. Spitzer, "Markov random fields and Gibbs ensembles," *American Math. Monthly*, Vol.78, pp.142-154, 1971.

[234] S.N. Srihari, R.M. Bozinovic, "A Multi-level perception approach to reading cursive script," *Artificial Intelligence*, Vol.33, pp.217-255, 1987.

[235] T. Steinherz, E. Rivlin, and N. Intrator, "Off-line cursive script word recognition - a survey," *Int. Jour. Document Analysis and Recognition*, Vol.2, No.2, pp.90-110, 1999.

[236] C.Y. Suen, C. Nadal, A. Mai, R. Legault, L. Lam, "Recognition of totally unconstrained handwritten numerals based on the concept of multiple experts," *Proc. Int. Workshop on Frontiers in Handwriting Recognition*, pp.131-140, 1990.

[237] B.J. Super, A.C. Bovik, "Three-dimensional orientation from texture using Gabor wavelets," *SPIE Conf. on Visual Communications and Image Processing*, pp.574-586, 1991.

[238] B.J. Super, A.C. Bovik, "Localized measurement of image fractal dimensional using Gabor filters," *Journal of Visual Communication and Image Representation*, Vol.2, No.2, pp.114-128, 1991.

[239] B.J. Super, A.C. Bovik, "Shape from texture by wavelet-based measurements of local spatial moments," *Proc. Int. Conf. on Computer Vision and Pattern Recognition*, pp.296-301, 1992.

[240] K.R. Tampi, S.S. Chetlur, "Segmentation of handwritten characters," *Proc. 8th Int. Joint Conf. on Pattern Recognition*, pp.684-686, 1986.

[241] I. Taylor and M.M. Taylor, *The Psychology of Reading*, Academic Press, 1983.

[242] S. Theodoridis and K. Koutroumbas, *Pattern Recognition*, Academic Press, London, UK, 1999.

[243] D.C. Tseng, H.P. Chiu, and J.C. Cheng, "Invariant handwritten Chinese character recognition using fuzzy ring data," *Image and Vision Computing*, Vol.14, pp.647-657, 1996.

[244] S. Tsujimoto, H. Asada, "Resolving ambiguity in segmenting touching characters," *Proc. 1st Int. Conf. on Document Analysis and Recognition*, pp.701-709, 1991.

[245] F. Ulupinar and G. Medioni, "Refining edges detected by a LoG operator," *Computer Vision, Graphics, and Image Processing*, Vol.51, pp.275-298, 1990.

[246] H. Urey, W.T. Rhodes, S.P. DeWeerth, and T.J. Drabik, "Optoelectronic image processor for multiresolution Gabor filtering," *Proceedings of IEEE ICASSP*, Vol.6, pp.3236-3239, May 1996.

[247] S.V. Vaseghi, "State duration modeling in hidden Markov models," *Signal Processing*, Vol.42, pp.31-41, 1995.

[248] A. Vinciarelli, "A Survey of off-line cursive word recognition," *Pattern Recognition*, Vol.35, No.7, pp.1433-1446, July, 2002.

[249] A.J. Viterbi, "Error bounds for convolutional codes and an asymptotically optimum decoding algorithm," *IEEE Trans. on Information Theory*, Vol.3, pp.260-269, April 1967.

[250] A.J. Viterbi and J.K. Omura, *Principles of Digital Communication and Coding*, McGraw-Hill, New York (1979).

[251] L. Wang and T. Pavlidis, "Direct gray-scale extraction of features for character recognition," *IEEE Trans. on Pattern Analysis and Machine Intelligence*, Vol.15, No.10, pp.1053-1067, October 1993.

[252] Z. Wimmer, S. Garcia-Salicetti, B. Dorizzi and P. Gallinari, "Off-line cursive word recognition with a hybrid neural-HMM system," in: *Advances in Document Image Analysis*, Eds, N.A. Murshed and F. Bortolozzi, Lecture Notes in Computer Science 1339, pp.249-260, 1997.

[253] M.K. Wright, R. Cipolla and P.J. Giblin, "Skeletonization using an extended Euclidean distance transform," *Image and Vision Computing*, Vol.13, No.5, pp.367-375, June 1995.

[254] T.P. Weldon, W.E. Higgins and D.F. Dunn, "Efficient Gabor filter design for texture segmentation," *Pattern Recognition*, Vol.29, No.12, pp.2005-2015, December 1996.

[255] Y. Xia, "Skeletonization via the realization of the fire front's propagation and extinction in digital binary shapes," *IEEE Trans. on Pattern Analysis and Machine Intelligence*, Vol.11, No.10, pp.1076-1086, October 1989.

[256] L. Xu, A. Krzyżak and C.Y. Suen, "Methods of combining multiple classifiers and their applications to handwriting recognition," *IEEE Trans. on Systems, Man, and Cybernetics*, Vol.22, No.3, pp.418-435, 1992.

[257] D. Yu and H. Yan, "An Efficient algorithm for smoothing, linearization and detection of structural feature points of binary image contours," *Pattern Recognition*, Vol.30, No.1, pp.57-60, 1997.

[258] J. Zerubia and R. Chellappa, "Mean field annealing using compound Gaussian Markov random fields for edge detection and image estimation," *IEEE Trans. on Neural Networks*, Vol.4, No.4, pp.703-709, July 1993.

[259] C. Zetzsche, T. Caelli, "Invariant pattern recognition using multiple filter image representations," *Computer Vision Graphics and Image Processing*, Vol.45, pp.251-262, 1989.

[260] Y.Y. Zhang and P.S.P. Wang, "A New parallel thinning methodology," *International Journal of Pattern Recognition and Artificial Intelligence*, Vol.8, No.5, pp.999-1011, 1994.

[261] L.A. Zadeh, "Fuzzy logic and soft computing: issues, contentions, and perspectives," *Proc. 3rd Inter. conf. Fuzzy Logic, Neural Nets and Soft Computing*, pp.1-2, Iizuka, Japan, 1994.

[262] J.M. Zurada, *Introduction to Artificial Neural Systems.* St. Paul: West, 1992.

Index

Druck: Strauss Offsetdruck, Mörlenbach
Verarbeitung: Schäffer, Grünstadt